PHalarope Books

PHalarope Books are designed specifically for the amateur naturalist. These volumes represent excellence in natural history publishing. Each book in the PHalarope series is based on a nature course or program at the college or adult education level or is sponsored by a museum or nature center. Each PHalarope Book reflects the author's teaching ability as well as writing ability.

BOOKS IN THE SERIES

SUBURBAN WILDLIFE
A Guide to the Common Animals
of Your Back Yard and Local Park
Richard Headstrom

SUBURBAN WILDFLOWERS
A Guide to the Common Wildflowers
of Your Back Yard and Local Park
Richard Headstrom

THE WILDLIFE OBSERVER'S GUIDEBOOK
Charles E. Roth, Massachusetts Audubon Society

THE AMATEUR NATURALIST'S HANDBOOK
A Classic Now Revised and Expanded—
to Make All the Outdoors Your Classroom
Vinson Brown

THE AMATEUR NATURALIST'S DIARY
Vinson Brown

AT THE SEA'S EDGE
An Introduction to Coastal Oceanography
for the Amateur Naturalist
William T. Fox, illustrated by Clare Walker Leslie

THE SEASIDE NATURALIST
A Guide to Nature Study at the Seashore
Deborah A. Coulombe

THE CURIOUS NATURALIST
John Mitchell and the Massachusetts Audubon Society

NATURE DRAWING
A Tool for Learning
Clare Walker Leslie

Jane Scott has taken botany and horticulture courses at Pennsylvania's famed Longwood Gardens. She has illustrated wildflower guides for the Delaware Nature Education Society and the Brandywine Conservancy, as well as a forthcoming book on flowers to be published by Prentice-Hall. She has also written two children's books.

BOTANY IN THE FIELD

An Introduction to Plant Communities for the Amateur Naturalist

Jane Scott

Illustrations by the Author

PRENTICE HALL PRESS • NEW YORK

Library of Congress Cataloging-in-Publication Data

Scott, Jane, 1931–
Botany in the field.

(PHalarope books)
"A Spectrum Book."
Bibliography: p.
Includes index.
1. Plant communities. 2. Botany. I. Title.
QK911.S36 1984 581.5'247 83–19180
ISBN 0-13-080292-1 (pbk.)

Published in 1986 by Prentice Hall Press
A Division of Simon & Schuster, Inc.
Gulf + Western Building
One Gulf + Western Plaza
New York, NY 10023

Originally published by Prentice-Hall, Inc.

Manufactured in the United States of America

10 9 8 7 6 5 4 3 2 1

First Prentice Hall Press Edition

CONTENTS

To Tony
who helped me all the way

PREFACE

This is a book for people who love plants: amateur naturalists in the literal, best sense of the word. If you like to hunt for spring wildflowers in the woods and summer ones in meadows, or if you wade exuberantly into bogs and salt marshes, oblivious to wet shoes and black mud, this book is for you.

Yet you will not find descriptions of every plant you see. This is not a field guide but an attempt to put plants in context. Putting a plant into context means to show it as a member of a community or a group of species that have adapted to the same habitat. Although some plants need the rich soil of woodlands, there are others that will push up through cracks in a sidewalk or hug the driest seaside dune. If you don't have to be told that the same species will not do all three, you already know something about plant communities.

There will always be plants in the landscape for they are among the most persistent of life forms. Yet some plants are very set in their ways. Having adapted to the requirements of their habitat over millions of years, they are destroyed by the slightest disturbance. Others are flagrant opportunists, putting down roots wherever they get the chance. The numerical relationship of one kind to the other tells us enlightening, but sometimes painful, tales about our past use of the land. We learn to read the landscape like a book.

The more chapters we read of this book, the more we come to value the original native plant communities, not only because they are the product of millions of years of evolution, but also because they clothe the land with an orderliness and beauty that enhances our sense of place, a quality so often lacking in this era of cheap mass production. Of course, it would be folly to suggest that we should never disturb the land. Civilization demands it. Yet we can learn to use land in a way that accommodates healthy and diversified plant communities. Such communities are the very basis of a healthy environment.

The knowledge and enthusiasm of innumerable botanists, past and present, have provided the basis of this book. Although it is impossible to list them all, my special thanks go to Dr. Donald Huttleston, taxonomist at Longwood Gardens, Kennett Square, Pennsylvania, for sharing his knowledge generally and especially on trips to the Great Smokies and the New Jersey Pine Barrens. I am also grateful to George Fenwick, chief of the Maryland Natural Heritage Program, who introduced me to a freshwater marsh on Maryland's Eastern Shore and also read some of the manuscript, and to Hal Bruce, curator of Plant Collections at Winterthur Gardens, and superlative writer, who led me through the Winterthur meadow. Norman H. Dill and Arthur O. Tucker, both of the Department of Agriculture and Natural Resources at Delaware State College, took me to see rare plant communities of the Coastal Plain. John and Mary Watkins helped me explore the oak woods of the Shenandoah and also contributed material on the pine barren communities of North Carolina. Ann R. Daudon, Nancy G. Frederick, and Elizabeth Sharp, all amateurs in the best sense of the word, lent me books on botany and taxonomy. Norman Wilder, executive director of the Delaware Nature Education Society, and Lorraine Fleming, research coordinator of the Society, both read and commented on parts of the manuscript. Enola Teeter, librarian at Longwood Gardens, helped me to find pertinent books and materials, and F. M. Mooberry inspired me to begin. If this book contains anything of value, it is due to the help and support of all these people.

Grateful acknowledgment is given to the following for granting permission to reprint material.

Not a tree
A plant, a leaf, a blossom, but contains
A folio volume. We may read, and read
And read again, and still find
something new

JAMES HURDIS (1763–1801)

I

THE LANDSCAPE

1

THE PLEASANT COMMUNITY OF PLANTS

Skunk Cabbage
Symplocarpus foetidus

Shadbush
Amelanchier spp.

In early spring, when flowering red maples, *Acer rubrum,* show a wash of crimson across the distant woods, when the hooded flowers of skunk cabbage, *Symplocarpus foetidus,* have already begun to fade but their kelly-green leaves brighten every swamp and stream bank, when a shadbush, *Amelanchier* spp., in bloom seems to drift like smoke among the oaks and pines, it is time to begin.

Among the special appeals of botany is the fact that plants, unlike animals, do not vanish at our approach. Yet they do move in time, shifting images with the revolving seasons like a slow-motion kaleidoscope, as bright fruits replace delicate flowers and the pervasive green of summer explodes into autumn's fire. In winter, every species of tree shows a distinctive silhouette, enhanced perhaps by caps of snow or ice gleaming in the sun, and dry beech leaves rustle like parchment against the pearl gray trunks. Down in the meadow we inspect the seed heads of grass and goldenrod and speculate on the promises of each round green rosette that hugs the frozen soil.

As spring progresses, we may find that certain wildflowers have disappeared from well-remembered sites and reappeared somewhere else. Individual plants cannot move, but whole populations are able to migrate in mystifying leaps, sometimes over long distances, thanks to that minute miracle: the seed.

The ferns, horsetails, and club mosses of today still reproduce by spores in the primitive two-stage manner of the first land plants of hundreds of millions of years ago. The mature plant releases spores from tiny capsules to blow on the wind. If they land on moist ground, they develop into the second stage of reproduction, a tiny plant called the prothallus containing the male and female organs. Like the sea plants from which it evolved, the prothallus has swimming sperm and is dependent on the presence of water, in this case merely a film, to accomplish successful fertilization.

The conifers came next on the evolutionary scene. They have male and female cones right on the tree, eliminating the uncertainties of an intermediate generation. Each cone has scales arranged spirally around a central axis that contain pollen sacs or ovules. Pollen from the male cones is dispersed by the wind to pollinate ovules in the larger female cones. Seeds then develop behind the scales: seeds that are able to wait out unfavorable conditions, independent of water, until the time is right for them to grow. That this has proved to be an efficient method of reproduction is demonstrated by the abundance of conifers still found on earth, yet the opportunities for seed dispersal pale beside those of the next evolutionary group: flowering plants.

Virginia Pine
Pinus virginiana

There were no insects on earth when the conifers evolved, but when flowering plants appeared, one hundred and thirty million years ago, they were plentiful. Conspicuous flowers are, in fact, an opportunistic attempt to entice and employ insects for purposes of pollination. As time went on both insects and flowers became more and more specialized and adapted to one anothers' needs. Insects were able to carry the pollen over distances not possible when fertilization depended on the capriciousness of the wind. The resulting seeds, too, had become specialized. Now they developed within the protection of an ovary and gradually became equipped with many ingenious methods of travel. Today many fly on the wind with tufts of hair for parachutes, while others, such as the seeds of maples and ashes, are equipped with sails like propeller blades and twirl through the air like tiny helicopters. Some seeds are so fine they simply blow about like dust; others hitch rides on the feet of wading birds. Delectable fruits often have indigestible seeds, a subtle but effective means of travel, whereas stick-tights and burrs simply cling to the coats of animals or humans.

Because all these seeds are sprinkled randomly

over the landscape, one would expect a random mixture of species to result, yet even the most cursory glance tells us this is not the case. Woods do not mix with marsh, and the plants found in a grassy meadow are not the same as those growing on a mountain slope.

The truth is that plants are cliquish. Similar species occur in similar groups in similar habitats all the globe, whether these habitats are contiguous or not. Each plant in a community arrives there independently, but rarely does a single species occupy a site to the exclusion of all others, with the possible exceptions of the red mangrove, *Rhizophora mangle,* in coastal Florida and the stands of giant reed, *Phragmites communis,* that have spread over disturbed marshes in the Northeast.

Habitat, then, is the key that determines plant communities, but what determines a habitat? Over a large geographical area, the answer is climate. Yet this is only part of the story, because numerous and diverse mini-habitats occur within broad areas of similar climate, depending on the available sunlight and water as well as the general character of the soil. In fact, several mini-habitats may occur on a single hillside with different plant species occupying the top, the bottom, and the northern and southern slopes.

As in human communities, individual species in a plant community both support and compete with one another. Nor is the habitat the same for all. The crown of a forest tree basks in bright sunlight, whereas the ferns at its base grow in dense shade. Yet without the tree to shelter them and its decaying leaves to provide humus for the soil, the woodland ferns could not grow. Thus, the increasing shade of a developing woods makes it possible for new species to migrate there, as the plants themselves gradually alter their own habitat.

In botanical terminology, a *stand* is a plant community that is homogeneous in all its layers, and an *ecotone* is the boundary between differing stands.

An ecotone may be as abrupt as a cliff or as gradual as a sloping hillside, depending on whether the plant species that dominate one community mingle with those of another before disappearing. Ecotones may exist in time as well, as one set of plant species is replaced by a succeeding group on the same site.

When botanists speak of the range of a plant, they mean the whole geographical area over which it is found. In field guides, the phrase "from Labrador to Georgia" usually means a plant that is widespread in the north but is found only high in the mountains farther south. Conversely, "Florida to New Jersey" means a southern species, abundant in the lowlands of the south that because of the moderating effect of the Gulf Stream has been able to migrate northward along the Coastal Plain. Ranges can be discontinuous, sometimes startlingly so. For instance, many genera of familiar North American plants, such as rhododendrons, witch hazel, sassafras, and magnolias, are also found in Asia. The genus of the tulip poplar, *Liriodendron tulipifera*, contains only two species in the entire world, one in the eastern United States and the other in eastern China! Botanists think both populations are survivors of a vast temperate forest that once covered the earth millions of years ago.

The rarity or commonness of a particular species refers to the distribution throughout its range. A plant that is thinly spread over a large range may not be classified as rare even though it is never plentiful at any one location, whereas a rare plant may be locally abundant but have a severely limited range. Plants become rare or even extinct for a variety of reasons. Most common, and most disturbing, is the destruction of habitat. Man has always destroyed habitat to improve his own condition, but in the twentieth century both his numbers and his mechanical efficiency have increased to such a degree that now, somewhere on the earth, a plant or animal disappears almost every day. Some of this is natural and beyond our power to stop, but all too much is

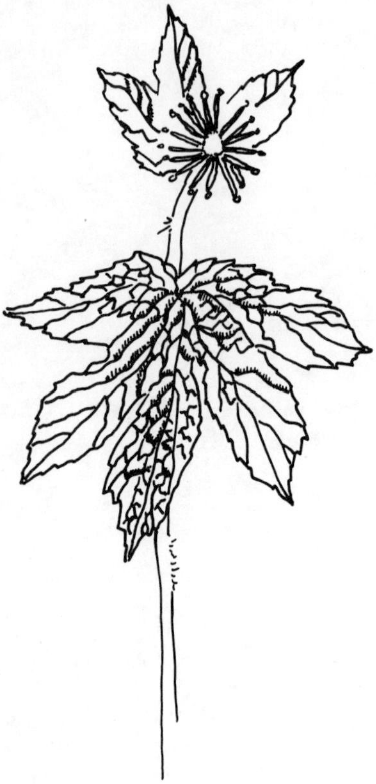

Goldenseal
Hydrastis canadensis

due to recklessness and ignorance. Overcollecting, especially by commercial growers, has also threatened the survival of certain plants, while others, such as ginseng, *Panax quinquefolius,* and goldenseal, *Hydrastis canadensis,* have been stripped from the woods because of alleged medicinal qualities. We are destroying valuable genetic diversity, which may indeed include the loss of potential medicines and food, in addition to diminishing the quality of life on this planet.

Habitats with unusual growing conditions contain plants that have evolved to accommodate those conditions and are ill adapted to other environments. If their habitat is destroyed, they will simply disappear. On the other hand, certain plants listed as rare on old plant lists may now be common because the habitats they prefer have become widespread. For example, prairie plants migrated eastward as the land opened. At the same time, a native species that cannot compete successfully with an invasive newcomer will decline and may eventually become rare or even extinct. It's also possible that a rare species has not finished migrating. Perhaps its method of seed dispersal is inefficient or areas of inhospitable habitat, too large to migrate across, divide its present location from other suitable environments.

We know that plants tend to migrate from the center to the perimeter of their range and that the farther they travel from home base, so to speak, the less tolerant they become of changing environmental conditions. Even so, the range of a southern plant may gradually creep northward during a period of mild winters, or fingers of forest may invade the prairie over a decade of wet summers, but eventually a fierce winter or a brutal drought will stop them. Thus, it is often extreme rather than average weather conditions that build the invisible "environmental fence" that marks the limit of a plant's range.

Of course, extremely cold winters or dry summers will affect the vigor of plants in the center of their range as well, often reducing the number of species found in a community as well as the health of individual plants. Large trees are rarely killed by one or two bad years, but their growth may be considerably slowed. Perennials may be affected in height and vigor, but annuals and biennials may completely disappear if but one year's crop of seeds fails to germinate.

For a plant population to maintain its numbers within a community, one seedling from each plant must reach maturity. It doesn't sound like very much considering the numbers that are produced every year, yet a large proportion of seeds are eaten by wildlife or fall on inhospitable ground. Of those that do manage to germinate, the majority succumb within the first year. Many woodland wildflowers, such as trillium or bloodroot, *Sanguinaria canadensis*, actually produce surprisingly few seeds.

Bloodroot
Sanguinaria canadensis

It is generally assumed that most North American plants have already migrated to the limits of their range. Yet many plants are successfully grown in gardens far to the north or south of their geographical homes. One can't help wondering if, given another thousand years, they will migrate to that latitude on their own.

Experiments in Louisiana's pitcher plant bogs have shown that natural hybridizing between species of pitcher plants tends to increase when the plants are under environmental pressure, and this may be true of other species as well. We know that hybridizing occurs frequently between species in the same genus, but because most fail to reproduce true to type they eventually disappear. Even if the new species is able to reproduce its own kind, it may simply fail to compete successfully with its own parents. Still, hybridizing is an important means by which plants evolve new species that are able to compete outside their original habitat, thus enlarging the range of the genus. Plants also mutate often, but again the new plant must be able to compete and to reproduce in kind if it is to survive. Two notable examples of spontaneous mutations which man has reproduced for his own use are the pink dogwood, *Cornus florida*, and the navel orange. The seeds of a pink dogwood may or may not produce pink offspring, and the navel orange, having no seeds at all, obviously cannot reproduce its own kind. Thus both plants must be reproduced vegetatively from cuttings or they would disappear.

Over the millennia, plants have adapted to their ranges in fascinating ways. For instance, the seeds of many northern plants will not sprout until they have gone through a period of freezing weather, thus automatically insuring the young seedlings the benefit of a full growing season before they must endure a northern winter. (Incidentally, the same mechanism prevents northern plants from extending their range into the south.)

Herbaceous plants of the temperate zone die to the ground in winter, a device that protects them from cold, breakage, and desiccation. Deciduous plants go dormant, dropping their leaves to prevent transpiration while the ground is frozen and water unavailable. Tree bark, which is made largely of cork, also protects trees from winter's drying winds. The needles of northern conifers are far more resistant to cold and drought than the leaves of deciduous trees, and because they remain on the trees all year, are able to carry on photosynthesis over a much longer period. Some continue until the temperature drops several degrees below freezing. Consequently, conifers are particularly well suited to high altitudes and northern latitudes, where the length of the growing season is far too brief for hardwood trees. Conifers also thrive in the Pacific Northwest, where 90 percent of the rainfall falls in the winter, leaving the summers far too dry for deciduous trees.

2
MAN AND THE LAND

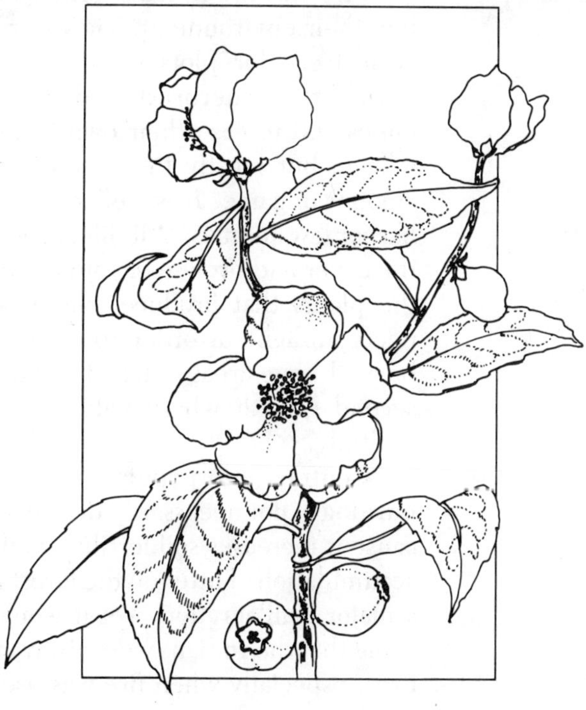

We usually assume that when the Europeans first arrived on this continent, a vast, unbroken forest covered all the land from the Atlantic coast west to the Mississippi River; a forest full of ancient trees with massive trunks and foliage that soared a hundred feet above a carpet of lush ferns and wildflowers. Such a forest certainly did exist; there are records of black walnuts six feet in diameter and hollow sycamores big enough to use as a barn or even shelter a frontier family until their cabin could be built. However, some historians claim that this forest could not have been as widespread nor as untouched as we suppose, because the Indians had been burning it for centuries.

Fire was the Indians' principal tool for controlling their environment. They burned the woods to clear trees for plots of maize and squash; they burned the underbrush to deny cover for their enemies and to ease their own travel from village to village; but mostly they used fire to increase the amount of game. It is well known that deep woods harbor few animals. Wildlife, especially deer, needs the cover and browse of young growth, and most of the plants that produce edible berries and seeds need sunlight in order to bloom. So the Indians burned large areas of forest to produce ecotones of second growth where wildlife could flourish and multiply.

Contrary to popular impression, controlled burning is not necessarily destructive to woods, because it increases soil fertility while preventing the accumulation of litter on the forest floor that can fuel a major conflagration. Yet it would be naive to assume that the Indians were always in control of their fires, especially when fire was used as a weapon of war.

Fire does alter the distribution of species, favoring some plants over others. For instance, while oaks and pines are known to be more common in a woods that has been frequently burned, there is also intriguing evidence that trees such as the rare southern yellowwood, *Cladrastis lutea*, may be fire dependent. This is not to say we should go out and burn the woods to increase the incidence of such species, but only that fire may play a part in the development of an ecosystem in ways that are not yet fully understood. Yellowwood, incidentally, is the single North American species of a genus also found in China and Japan. Its flowers are white, fragrant, and pendulous, and it blooms in June.

Yellowwood
Cladrastis lutea

The Europeans were afraid of the forest, which is not surprising considering that they were faced with the need to sustain themselves on a vast empty continent that was completely isolated from the support of settled civilization. They soon copied the Indians and began burning the woods to clear space for settlements and farms. What they didn't burn, they cut down.

England, as the mother country, believed it the duty of her colonies to supply whatever raw materials she required, and what she principally required at that time was lumber, particularly oak timbers and pine masts for the Royal Navy. That her own forests had been depleted long since and her admiralty forced to import wood from the Baltic coast did not, unfortunately, inspire the British to practice conservation in the New World. The crown sent emissaries deep into the forests of New England to mark the best trees by hammering an iron brand in the shape of a crown and anchor into the bark. This mark may still be seen on an occasional Maine giant that managed to escape the ax.

By the eighteenth century, the colonists were building sturdy, often elegant houses as well as creating beautiful furniture from the abundant supply of wood. The cutting of the forest continued throughout the Revolution and beyond, and as the lumbermen moved ever westward in search of virgin territory, they became folk heroes of Paul Bunyan-like proportions. Everywhere they went the new settlers viewed the woods as their enemy, not

Blue Lobelia
Lobelia siphilitica

Spring Beauty
Claytonia virginica

only because it interfered with farming, but because it also provided cover for dreaded Indian attacks.

It must have seemed as if no one cared about the destruction of the trees, yet this was also a time of meticulous and enthusiastic interest in natural history. Plants were the principal source of medicine, and Europe was eager to find cures for such widespread complaints as "wasting fever," syphilis, and "pox." Sassafras, *S. albidum,* familiar to us as an attractive small tree of old fields and fence rows, was singled out as a medicine tree as early as 1574 by Nicholas Monardes of Seville. In a pamphlet entitled "Joyful Newes out of the Newe Founde World," he recommended sassafras as a remedy for malaria as well as for a variety of other complaints. The belief grew all over Europe that sassafras was a magic cure of the Indians and it became as valuable a commodity to the early settlers of Virginia as tobacco. Whole expeditions were sent from England for the express purpose of gathering sassafras root. As time went on, many other American plants also became sought after, among them, blue lobelia, *Lobelia siphilitica,* as a cure for syphilis and Seneca snakeroot, *Polygala senega,* as a protection against snakebite.

It wasn't long before a brisk traffic in seeds and plants was crossing the North Atlantic. In 1678, the Temple House Botany Club, together with the Bishop of London (an enthusiastic botanist himself, named Henry Compton), sent a young man named John Banister to collect plants in Virginia. By the early eighteenth century, two native-born Virginians, John Mitchell and John Clayton, were also describing the local flora. Their names are immortalized in two of the plants they found, *Mitchella repens,* the partridge berry, and *Claytonia virginica,* the spring beauty. By 1733 Mark Catesby had finished his beautifully illustrated classic, *A Natural History of Carolina, Florida and the Bahama Islands,* and further north John Bartram had begun a collection of plants on a small farm outside Philadelphia. Born in 1699, John Bartram had an intense lifelong interest in botany. Although his formal education was slight, as a young man he had bought himself a botany text and learned Latin so he could read it. His first box of American seeds and plants was sent to London in 1734 to a Quaker merchant named Peter Collinson with whom he was to correspond for the next thirty years. In 1736 Collinson formed a group of interested botanists, each of whom paid five guineas in return for the seeds of a hundred North American trees and shrubs, each minutely described by Bartram. Many of these became prized ornaments in English gardens. By midcentury, Bartram was respected to the highest levels of European society and in 1765 even became official botanist to the king of England. He continued to make trips into the American wilderness, usually traveling on a horse laden with leather bags full of his precious specimens. It was during a collecting trip to Georgia in 1770 that Bartram discovered a lovely small tree along the banks of the Alatamaha River which he named *Franklinia alatamaha*, in honor of Benjamin Franklin. This tree was never found in the wild again after 1803. The cuttings taken by John Bartram and later by his son William Bartram are credited with saving it from extinction, although some modern ecologists fear that their zealous collecting may in fact have contributed to its demise. At any rate, all of the Franklinias grown in modern gardens are descendants of their trees.

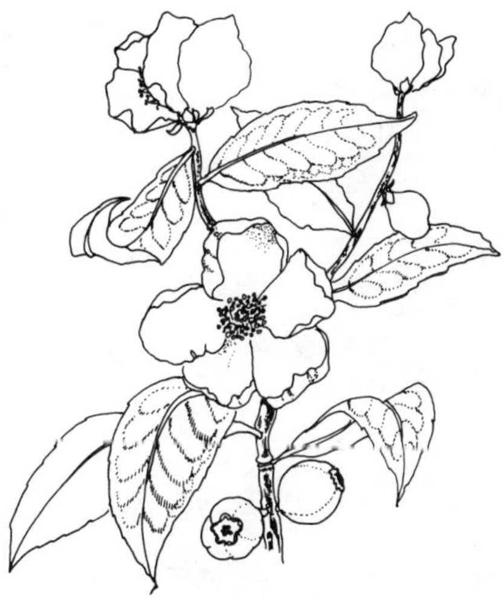

Franklinia
Franklinia alatamaha

William Bartram inherited his father's love for plants and combined it with a natural talent for writing. The colorful descriptions in his book *Travels*, published in 1791 after a four-year and five-thousand-mile trip through the southeastern United States, are credited with inspiring Samuel Coleridge to write his famous poem "Kubla Khan."

All these early botanists were deeply indebted to the great Swedish naturalist Carl Linnaeus. Linneaus was born Karl von Linne in South Rashult, Sweden. His father, a parish curate, wanted him to study for the ministry, but his son's interest in plants was so intense that the senior von Linne was finally persuaded to send him to medical school, for at that time a knowledge of botany was the principal requirement for the practice of medicine. While in medical school, young Karl started a botanical garden, writing meticulous descriptions of all the plants. Because he wrote in Latin, he became known as Carolus Linnaeus. His new system for the scientific classification of plants, first published in *Species Plantarum* in 1753, was based on sexual characteristics. The number of stamens and pistils, the plant's male and female organs, determined the class and order to which it belonged (see Chapter Nine). In addition, each plant was assigned one generic and one specific name, thus doing away with the long and complicated descriptions that had been previously used. Linnaeus's system was both simpler and more coherent than any that had preceded it, making it possible for botanists to describe new plants accurately and also to communicate clearly with each other for the first time in history. Eventually, as knowledge of biology grew, Linnaeus's system of classifying plants became constricting and was modified by later botanists, yet the simplicity and elegance of his binomial system of nomenclature continue to serve us well.

Linneaus himself was deeply interested in the flora and fauna of the New World and had sent his own envoy to America in 1748. This was Peter Kalm, the man for whom *Kalmia*, or laurel, is named. His book *Travels in North America*, also published in 1753, was widely sought after in a Europe long devoid of wilderness, where the populated land had been encumbered by feudal restrictions for centuries. The Europeans thirsted for news of this vast new continent where individual property could be had for the taking, and strange new varieties of plants and animals were everywhere.

The European nations never recognized any political claim of the Indians to land in America, but immediately took title for themselves upon discovery and settlement. The English crown commonly granted charters from afar to colonial governors and organizations such as the Massachusetts Bay Company, who then distributed land to individuals in an effort to encourage further settlement. Thus, by the early eighteenth century most of British America was populated by independent farmers living on a few hundred acres cleared from the forest. Here is Peter Kalm's description of farming practices in early Pennsylvania:

> The Europeans coming to America found a rich, fine soil . . . lying as loose between the trees as the best bed in a garden. They had nothing to do but cut down the wood, put it up in heaps and clear the leaves away. . . . This easy method of getting a rich crop has spoiled the settlers, and induced them to adopt the same method of agriculture as the Indians; that is, to sow uncultivated grounds as long as they will produce a crop without manuring, but to turn them into pastures as soon as they can bear no more, and to take on new spots of ground, covered since ancient times with woods, . . . [and] spared by the hatchet since Creation.

The phrase "nothing to do but cut down the wood" seems to oversimplify the problem. In fact, they usually girdled the trees and let them die or chopped halfway into the trunk and let the wind do the rest. Eventually, when the dead wood was dry enough, they burned it.

When the Revolution was over and the new Constitution ratified, all the land west of the Appalachians that had been previously claimed by the states was turned over to the new federal government. It was then sold to individuals to help pay off the war debt, a practice that instigated a surge of agriculture that was to sweep over the country for the next hundred years. More and more forests fell to ax and fire, and the wild prairie sod was severed by the plow. By the midnineteenth century, when the Industrial Revolution was causing a shift of the population from the farms to urban areas, trees were felled for charcoal to fuel the smelting furnaces of the burgeoning iron industry.

As midwestern farms grew larger, their crops became geared to commercial markets, while the East shifted to sheep and dairy herds. In 1909, when James Taylor, headmaster of a school in Vermont, conceived the idea of the Long Trail, a foot-path that still stretches from Williamstown, Mass. over the crest of the Green Mountains to the Canadian border, the slopes of many of those mountains were sheep pastures, grassy and bare of trees. Even today it is possible to find the remains of stone walls that once fenced in the flocks near the crests of the ridges.

Today the Atlantic Coastal Plain is still heavily farmed, but many hills in New England and the Piedmont are really too steep for efficient agriculture, and as time went on the farmers found they could not compete with the huge crops that flooded in from the flat land of the Midwest, nor with the huge herds of the western range. Many moved west themselves, while others simply gave up farming, leaving their fields to succession. Today those fields are well into second growth, but they are being cleared again for housing developments and shopping malls.

Many eastern farms were abandoned during the Depression of the 1930s, when farmers could not pay off their debts. Near my house in southeastern Pennsylvania, there is a stand of tulip poplar, *Liriodendron tulipifera*, a species that acts as a pioneer hardwood in this area. Still visible on the woods' floor are faint but regular corrugations: they are the remains of furrows left by the plow. Because that last crop was never planted, an immediate change took place in the field's environment: a change that allowed the process of plant succession to begin.

3 PLANT SUCCESSION

Given seeds and time, a field of plowed earth will turn into a towering forest: an elegant example of nature's healing force that botanists call plant succession. The complex orchestration of species, each stage in its turn preparing the way for the next, begins the moment a piece of land is released from human domination. There are many variations on the theme, depending on the exposure of the site and the general characteristics of the soil, but the underlying rhythm is so predictable that it is often possible to tell within a few years when a particular field was abandoned.

At present most of the land in the East which is not plowed, mowed, grazed, paved, or built upon is in some stage of succession. The only exceptions are those places where the new growth has already reached *climax vegetation,* a botanical term describing the final diversified plant population that, in theory at least, will maintain itself indefinitely. In reality, however, some disturbance is always occurring. Trees are felled, if not by man, then by lightning, fire, or disease. As sunlight pours into the resulting clearing, the plants of an earlier stage reappear and the rhythm of succession reasserts itself. Even under the best of circumstances it will take another century for those woods now in climax to resemble the ancient forest that originally existed on this continent.

Daisy Fleabane
Erigeron annuus

The seeds of most of the plant sequences that will cover an abandoned field in the years to come are present at the moment of abandonment, but they do not all germinate at once or under the same conditions. Many will need a period of winter cold before they will sprout. Some herbaceous plants, a term that includes grasses, sedges, and rushes, as well as the forbs we usually think of as flowering plants, are annuals. They sprout, set seed, and die within a single growing season. Some are winter annuals, with seeds that germinate immediately upon ripening in late summer, spend the winter as ground-hugging rosettes, and bloom and die the following spring. Biennials go through an entire year as rosettes. The flower appears during the second summer, and the plant dies after producing seed. Perennials have hardy roots. They go on blooming and setting seed for years until something causes them to weaken and die. We usually think of perennials as herbaceous plants that die to the ground every winter, but shrubs and trees have hardy roots as well and could by this definition be classified as perennials.

Annuals, such as the notorious crab grass, *Digitaria sanguinalis,* will clothe the bare dirt of a plowed field almost immediately. Rosettes of winter annuals will soon follow, such as daisy fleabane, *Erigeron annuus,* horseweed, *E. canadensis,* and those members of the mustard family, genus *Brassica,* whose flowers will paint the hillsides yellow the following April. As the next spring fades into summer, annuals such as ragweed, *Ambrosia* spp., and Queen Anne's lace, *Daucus carota,* will appear. Certain fast-growing perennials, such as dandelion, *Taraxacum officinale,* and black-eyed Susan, *Rudbeckia hirta,* may flower during the first summer after abandonment, but goldenrod, *Solidago* spp., butterfly weed, *Asclepias tuberosa* and yarrow, *Achillea millefolium,* will usually need another year before reaching blooming size.

Field Mustard
Brassica rapa

Many of these plants of early succession are aliens; that is, they are natives of Europe or Asia and were brought here by man. They are the weeds that are quick to cover bare ground and interfere with our crops and gardens. It is estimated that alien species now make up as much as 18 percent of the total plant populations of the northeastern states. Most are sun lovers that when away from the limitations of their native environment thrive to the point of being invasive. Many arrived with the earliest settlers, their seeds hidden among hay brought for livestock or mixed in among seeds brought for garden crops. Others came in ship ballast or in the bottom of pack-

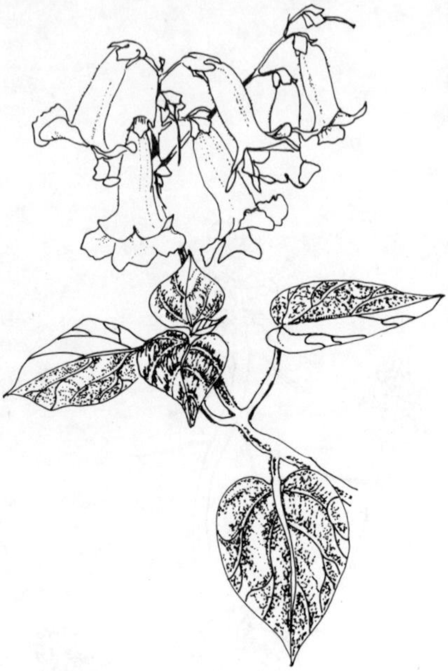

Empress Tree
Paulownia tomentosa

ing boxes. Some, such as the seed pods of the empress tree, *Paulownia tomentosa*, were used as packing material for fine porcelain, much as styrofoam peanuts are today. The wood of the empress tree is highly valued in Japan for dowry chests, but its population there has been declining. Consequently the Japanese have been recently buying *Paulownia* wood from the United States, completing the circle begun so many years ago. In more modern times, alien plants often come as ornamentals that escape into the environment because their seeds are attractive to birds.

By the second or third year our abandoned field is dominated by goldenrod and asters, *Aster* spp. Woody plants are now present as well and by the fifth year will take their turn as dominants. Then the field will become a dense thicket of blackberries, *Rubus* spp., sumac, *Rhus* spp., and two alien vines: Japanese honeysuckle, *Lonicera japonica*, and oriental bittersweet, *Celastrus orbiculatus*. Originally introduced as garden plants, both are rampant growers that twine over other vegetation, choking out woody plants and smothering herbaceous ones. They are so pervasive that many people don't realize that there are native species of both. Trumpet honeysuckle, *Lonicera sempervirens*, is a twining vine with red and yellow tubelike flowers, and *Celastrus scandens* is the American species of bittersweet. In both, the flowers are borne only at the ends of the stems, in contrast to the Japanese varieties, which have flowers in the leaf axils. Neither American species is invasive.

Multiflora Rose
Rosa multiflora

Farther south, kudzu, *Pueraria lobata*, another native of Japan, has draped itself over millions of acres, growing as much as a foot a day (see Chapter 6). A relative newcomer to this list of botanical villains is the multiflora rose, *Rosa multiflora*, a plant customarily used as the rootstock for grafting hybrid tea roses. A few years ago, an enterprising nurseryman advertised it as a "living fence," a plant so dense and prickly that it would contain pets and livestock.

When it was planted in backyards and along highways, it proved so attractive to birds that it is credited with the northern spread of cardinals and mockingbirds. Birds, in return, have spread it into every field and byway. All these intruders are unaffected by mowing, which only stimulates them to further growth, and can become so rampant in open fields that they slow natural succession. Eventually, however, pioneer trees such as dogwood, *Cornus florida,* red cedar, *Juniperus virginiana,* sassafras, *Sassafras albidum,* and wild cherry, *Prunus* spp., will succeed in poking through the tangle. When they are joined by tulip poplar, *Liriodendron tulipifera,* and red maple, *Acer rubrum,* the resulting shade will weaken the invaders.

Broomsedge
Andropogon virginicus

Often areas of broomsedge, *Andropogon virginicus,* will come in among the goldenrod. This is a tuft grass with densely fibrous roots that spreads in nearly unbroken swathes across some old fields. The only other plants in a field of broomsedge may be a few red cedar, hawthorne, *Crataegus,* or dogwood. It used to be thought that the soil in such areas was too poor to support other vegetation, but recent research has shown that *Andropogon* is alleopathic, that is, its roots exude a chemical which inhibits other plant growth. Consequently, a broomsedge meadow may persist for many years, sometimes surrounded by woods well into the pioneer hardwood stage.

By the end of a decade our field is well on its way to becoming a woods. As the trees begin to shade the field in summer, the early weedy plants give way. As decaying leaves increase the humus content of the soil, the tulip poplars and maples grow into stately trees, perhaps surrounding a dying red cedar straining toward the light. The dogwood takes its place as a subcanopy tree, joined by shrubs such as mapleleaf viburnum, *Viburnum acerifolium,* spicebush, *Lindera benzoin,* and witch hazel, *Hamamelis virginiana.* Ferns appear, and so do all the delicate wildflowers known as "spring ephemerals." Squirrels

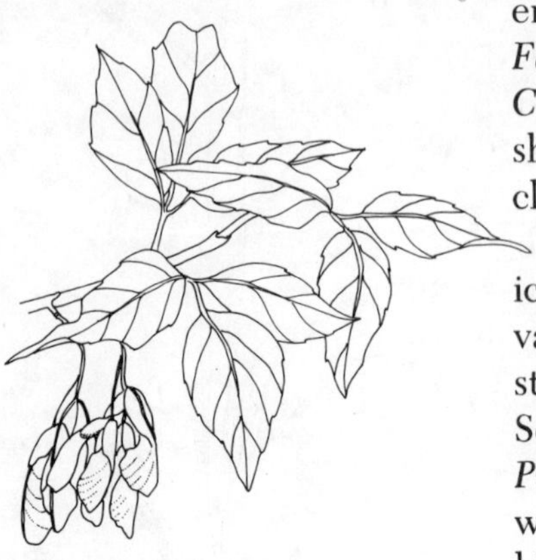

Box Elder
Acer negundo

entering the young woods bury the nuts of beech, *Fagus grandiflora*, oaks, *Quercus* spp. and hickory, *Carya* spp. The resulting seedlings grow well in the shade and are the species that will dominate the climax forest.

The successional stages I have described are typical of an old field sequence in southeastern Pennsylvania. In other areas and under different circumstances the species may differ. For instance, in the South where the soil is dry and sandy, Virginia pine, *Pinus virginiana*, and pitch pine, *P. rigida*, are woody pioneers, often followed by oaks such as blackjack oak, *Quercus marilandica*, and chestnut oak, *Q. prinus*. Dry soils slow plant growth wherever they occur and the rate of succession in such places is often slowed as well. Replacement occurs much faster on flood plains and bottom lands, where the soil is always moist. Here the early succession may be dominated by black willow, *Salix nigra*, box elder, *Acer negundo*, silver maple, *A. saccharinum*, and the climax by red maple, sycamore, *Platanus occidentalis*, swamp white oak, *Q. bicolor*, and black walnut, *Juglans nigra*.

The pioneer hardwoods of interior New England are usually birch, *Betula* spp., and white pine, *Pinus strobus*, followed by oaks and sugar maples, *A. saccharum*, with an eventual climax of hemlock, *Tsuga canadensis*, and beech, although much depends on the slope and exposure of the woods.

Particular species may vary, but a good rule of thumb is that the early successional stages everywhere are made up of sun-loving plants, a large proportion of which are aliens, with seeds that will not germinate in the shade. This is also true of early trees: young white pines cannot grow in the shade of old ones, for instance. Later successional stages are shade-loving and the majority are also indigenous to eastern North America.

Between fifty and one hundred years after the field was abandoned, the climax vegetation is in place, but it will take another century before it will resemble once again the optimum North American deciduous forest. Throughout history and on every continent, human beings have destroyed the woods to make room for crops and herds. In many places they have destroyed the land as well, exhausting the fertility of the soil to the point of creating deserts where none existed before. Now that we are nearing the end of the twentieth century, expanding populations worldwide are putting intense pressures on ecosystems everywhere, but in spite of past abuse, our familiar eastern woods are still viable enough to contain a variety and abundance of plants unsurpassed anywhere. One hopes that we will have the maturity to allow at least some of it to return, undisturbed.

4
PLANT PROVINCES

Of course, succession does not always end in forest. Bands of differing vegetation swirl across the North American map like marbled endpapers in an old first edition. Circling the globe just south of the polar icecap is an expanse of rust and green land known as the Arctic Tundra. Beneath it is a band of deep forest green that sweeps in a giant parabola from southern Alaska across Canada, dipping over the U.S. border into the northern edge of Michigan, New York, and New England. This is the taiga, or Northern Coniferous Forest. It is mottled with sky blue lakes and moss green sphagnum bogs, called muskegs by the Indians. Isolated drips of this forest green flow down the spine of the Appalachians as far south as western North Carolina and eastern Tennessee. A second coniferous band, called the Western Evergreen Forest, streams like a moire ribbon of differing tree associations from coastal Alaska into Mexico.

The eastern half of the United States is overlaid by a rich mosaic pattern known as the Eastern Deciduous Forest. It covers most of New England and the Piedmont country of the Atlantic states and spreads across the Appalachians to the Ohio River valley. It is the largest province within our borders as well as the most diverse, and its color varies with the seasons. Yellow-green in spring, it darkens to an older green in summer, exploding into the colors of fire every autumn before settling into an iridescent gray for winter.

South and east of the deciduous forest in the Coastal Plain. This is tidewater country, characterized by salt marshes, slow-moving rivers, pine forests, and cypress swamps. Not long ago, geologically speaking, it was the bottom of a shallow sea. The province includes much of what we still think of as the plantation country of the Old South, but it also extends around the base of the Appalachians into eastern Texas and up to the Mississippi River valley into southern Illinois.

Coating the tip of Florida is a piece of true Tropical Forest that also includes the Caribbean islands, coastal Mexico, and Central America. Members of the palm family, *Palmae*, are typical of this province. Seven genera are native to Florida, although they do not include the coconut palm, *Cocos nucifera*, nor the date palm, *Phoenix dactylifera*, as both are introduced species. The ubiquitous red mangrove, *Rhizophora mangle*, an aggressive colonizer of Florida's estuaries and lagoons, is typical of tropical coastal vegetation. It was widely scorned as a pest tree until recent studies identified it as an important link in the complex food chain of the south Florida coast.

West of the Appalachian mountains, the eastern forest reaches in long, fingerlike peninsulas into the green and ocher stain that is called the Grasslands Province. These great prairies of the central plains are generally divided into the tall grass prairie of the Midwest, where rain is comparatively plentiful, and the short grass prairies of the Far West.

West of the prairies is the "cold desert" of the Great Basin province of Utah and Nevada, and to the southwest the "hot desert" of the Sonoran Province of New Mexico and Arizona. Along the southern coast of California is the Chaparral Province or "hard-leaved forest," a type of dense scrub which develops in regions of mild, wet winters and very dry summers.

Of course, my map is obsolete. Two centuries of intensive civilization have reduced the remaining bits of native vegetation to areas that are few and far between. While a detailed examination of every community in every province is beyond the scope of one book, it is important to realize that they are plant communities on a grand scale, and that many of the same factors that determine their boundaries also determine the boundaries of smaller communities within them.

What are these factors? The simple answer is water, climate, soil conditions, and fire. Water, however, is not simply the amount of annual rainfall, for high temperatures also mean high evaporation rates, whereas at cool temperatures rainwater and snow are slow to disappear. Topography, too, influences drainage and the amount of water available to plants. It also makes a difference when the rain falls. Is it evenly divided throughout the year, or does most fall in one season? The term *climate* encompasses the annual rainfall and the range of seasonal temperatures plus the number of frost-free days (the length of the growing season). The varying hours of daylight and darkness are also important, for their shifting relationship is the signal that stimulates many plants to grow and bloom.

Soils everywhere are profoundly influenced by the underlying rock, but the size of the soil particles and the presence or absence of organic matter are also important. All of these affect the drainage as well as the pH factor, the relative acidity or alkalinity of the soil. It is the pH factor that affects the action of the soil bacteria that change nutrients into a form that is usable by plants. Fire, as we have seen, is the wild card. Its subtle effect on plant communities has been appreciated for relatively few years.

Consider the Arctic Tundra, where conditions are extreme. The annual rainfall is light, only about twelve inches a year, including snow. Yet only the top few inches of soil ever thaw. The subsoil is locked in permafrost, impeding drainage, and the low temperatures mean the rate of evaporation is slow. Consequently, the soil is both cold and saturated. Light is almost continuous during the brief arctic summer, but in winter the sun barely clears the horizon and the temperature averages zero or below. No trees grow on the tundra. Not only are their roots blocked by the permafrost, but trees must have at least eight weeks of temperatures above fifty degrees Fahrenheit in order to bloom and set seed. Tundra vegetation, therefore, is mainly moisture-loving mosses, grasses, and sedges, although there are low-growing ericaceous shrubs, dwarf willows, and some herbaceous plants as well.

Many of these burst into glorious bloom during the brief arctic summer.

It is an intriguing fact that in the temperate zone one can travel to northern climates by climbing mountains, and bits of Alpine Tundra can be found on the highest peaks of northern New England. Alpine Tundra, however, differs from true Arctic Tundra in a number of ways. In the mountains the summer days are shorter and the growing season longer. Snowfall is heavy and slow to melt. It is this, not permafrost, that inhibits the growth of trees, because the ground is never free of snow long enough for seeds to germinate and seedlings to develop. Moreover, drainage is usually good in the mountains, resulting in fewer mosses and more flowering plants.

Tundra rainfall, scarce as it is, stays. In the desert, what rainfall there is quickly drains away through the porous sandy soil and evaporates in the hot summer sun. Therefore, desiccation, or the excessive drying of their tissues, is the chief enemy of desert plants. There are also wide diurnal swings of temperature, often reaching well above a hundred degrees on a summer day, but dropping as much as forty degrees at night. In the so-called "cold desert," where winter frosts are not uncommon, sagebrush, *Artemisia tridentata*, is the dominant plant. "Hot desert" plants have fat, succulent stems and spiny leaves. They are often coated with a waxy surface to further protect them from desiccation. In the hot deserts of the Sonoran Province, cacti are abundant, as are agaves and yuccas. The desert is also famous for its myriads of bright annuals. During the brief periods of rain, they seem to spring from the earth to grow, bloom, set seed, and die within a few short weeks. Annuals are common in hot, dry climates, but they decrease in abundance as one travels north because cold soils are not conducive to quick seed germination.

As we shall see, our eastern sand dunes have

many desert characteristics. Even though rain is plentiful along the dunes and barrier beaches, the sand is so porous that water is not available to plants for long, and the hot sun and ocean winds dry their tissues.

Grasslands tend to develop in the interior of large land masses, where the temperature variation is extreme and prolonged droughts are frequent. They, like deserts, are often in the "rain shadow" of high mountain ranges. As moisture-laden clouds sweep in from the ocean, they are forced over the heights. The colder air causes the water to condense and fall on the peaks as rain or snow. Very little is left for the land on the other side.

Prairie winters are cold and windy, but the buds of grass lie safely dormant at or below ground level, well mulched by the thatch of previous years. They spurt upward in the warm spring rains, but can quickly revert to dormancy during the punishing droughts of summer. Prairie soil is more alkaline than that of the eastern woods, especially in the high plains of the western short grass prairie.

Grass evolved relatively late, probably in Eocene times, about sixty million years ago. Its appearance made possible the vast herds of bison and other herbivores as well as the busy populations of prairie dogs that once dotted the plains. Prairie dogs and bison, in turn, helped keep the grassland open. So did the Indians, who regularly set it afire to herd bison into circles for slaughter. Their fires, as well as those ignited by summer lightning, blew eastward on the prevailing winds, killing the woods and spreading the prairie into western New York and Pennsylvania. Now both herds and Indians have disappeared, the dense prairie sod has been cut by the plow, and the woods have spread west once again.

The dense coniferous forest crosses southern Canada and dips into our northernmost states, clothing a land of low relief. Northern soils are sandy and acid under a deep layer of slowly decaying litter.

They are called "podzols," a Russian word for ash-gray soil, and are easily recognized by a layer of white sand that is left near the surface as humic acids percolate slowly downward. Podzol soils are unsuited to agriculture, and that fact, plus the short northern growing season, have meant that, except for logging, boreal forests have been relatively undisturbed. Now, however, they are being seriously threatened by acid rain, which substantially lowers the pH of the soil, interfering with the nutrients available to the trees. This weakens them and makes them especially vulnerable to insect damage and disease.

Northern winters are long and cold, but again the rainfall is relatively sparse, about fifteen inches a year. The small, hard needles of coniferous trees resist desiccation, enabling them to survive where the evaporation rate is low. Further south, grassland would develop under the same conditions. Familiar spring wildflowers of the north woods are the goldthread, *Coptis groenlandica*, starflower, *Trientalis borealis*, and wood sorrel, *Oxalis montana*. The predominant tree species is the white spruce, *Picea glauca*. It is joined in the West by lodgepole pine, *Pinus contorta*, and in the East by balsam fir, *Abies balsamea*. Paper birch, *Betula papyrifera*, is everywhere, its gleaming white trunks the grace notes that enliven the somber music of the evergreens. Tamarack, *Larix laricina*, and black spruce, *Picea mariana*, circle the frequent sphagnum bogs, while jack pine, *Pinus banksiana*, and aspen, *Populus tremuloides*, are the first to appear in burned-over areas.

Starflower
Trientalis borealis

As mentioned, bits of this boreal forest extend far to the south along the spine of the Appalachians, usually above five thousand feet. These patches are remnants of the time when the northern forest retreated southward before the ice sheets, and coniferous trees covered the entire landscape of the East. As the climate warmed, deciduous forest gradually replaced the evergreens at low altitudes, isolating the conifers high in the mountains. Over the ages, new species evolved so that now the dominant southern conifers are the red spruce, *Picea rubens*, and the fraser fir, *Abies fraseri*. These trees grow both higher and denser than their northern cousins because rainfall is plentiful in the southern mountains—up to one hundred inches a year in the Smokies.

All of the land now covered by the northern forest was buried repeatedly by glaciers, vast continental ice sheets that crept southward four times during the last million years. In North America, they spread from two centers: the Laurentic sheet in the East and the Cordilleran sheet in the West. The last advance was called the Wisconsin glacier. It covered the Great Lakes and all the adjacent land eastward to New York City, retreating about twelve thousand years ago.

A glacier forms when the differential between the winter and summer temperatures is too small for the snow to melt. Year after year the snowfall builds, eventually compressing into an enormous mound of ice. The immense weight of the ice causes enough friction to melt the leading edge. As the water flows out and refreezes, the glacier begins to move. The alternate freezing and thawing also allow melt water to penetrate cracks in the surrounding rock, cracking off large chunks that are sometimes transported great distances on the back of the ice.

The glacier advanced across the landscape like a monumental plow, killing all vegetation and moving vast piles of earth before it. These materials, called moraines, were left behind when the ice finally melted, forming islands and land masses where none had existed before. The glaciers diverted the courses of rivers, carved mountains, mounded hills, and altered the level of the sea. The crushing weight of the ice actually compressed the land surface, which then sprang back unevenly, creating depressions of poorly drained land. Isolated blocks of slow-melting ice sometimes remained behind while deposits left by the retreating glacier piled up around them. When these blocks melted in their turn, they formed "kettle holes," which filled with melt water, forming the deep, clear glacier lakes of the North.

The glacier forced the vegetation to move southward ahead of it. Because the ancient Appalachian ridge runs north and south, the plants had room to retreat and were thus saved from the fate of numbers of European flora, lost forever in the east-west ridges of the Alps because the more tender species could not migrate over the top. This is one reason why the diversity of native European plant material is sparse compared to ours. Centuries of human occupation have further reduced the numbers.

As the climate warmed the deciduous trees began migrating north once again, a process that still continues. Deciduous trees cannot survive in the fierce, dry cold of the northern woods. They need about forty inches of rainfall spread evenly throughout the year. Even so, they drop their leaves in winter when the ground water is frozen and protect their incipient buds by a waterproof covering called a cuticle.

Today the Eastern Deciduous Forest covers most of the Appalachians and their foothills from New England to Georgia. It is hard to imagine just how old the Appalachian land mass really is. Geologists say as much as five hundred million years. If so, it was there before life crawled out of the sea, a sea that then washed over the central part of this continent. Three hundred and fifty million years ago, the New England mountains pushed up at the northern edge of Appalachia and the first lichens and algae covered the earth.

During the next one hundred million years, the seas withdrew, leaving vast swamps in which enormous plants called *Arthrophyta* flourished. The field horsetail, *Equisetum arvense*, a common plant

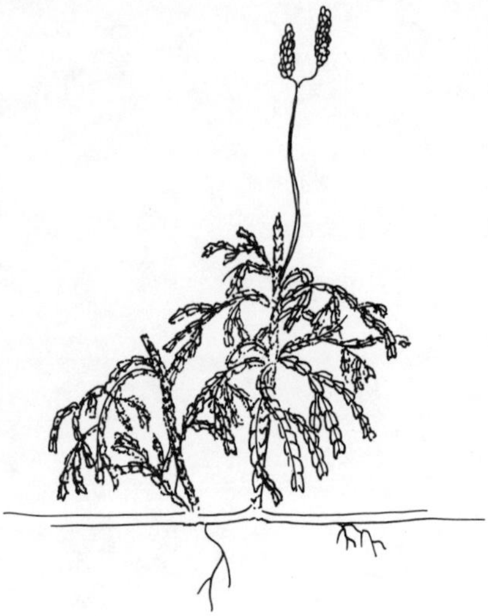

Running Pine
Lycopodium complanatum

of railroad embankments and wet, gravelly soils is one of its descendants. *Lepidophyta*, or scale trees, were also prevalent in prehistoric times. They have evolved downward, so to speak, into our running ground pines, *Lycopodium* spp. All these plants were prevalent when the dinosaurs ruled the earth. By the end of their time, one hundred and thirty million years ago, evolution took one of its startling leaps forward. Insects appeared, then flowering plants, and finally mammals. All three were to inherit the earth.

At that time the climate was apparently milder than it is today, and tropical forests advanced as far north as the state of Washington. During the next fifty million years, the western mountains, which had first appeared in Jurassic times, one hundred and fifty-five million years ago, rose to new heights and the climate cooled at high altitudes, causing a rain shadow on the eastern side of the range. The tropical forest was forced to retreat southward. During the Miocene period, twenty-eight million years ago, a temperate forest moved southward from Alaska and northern Canada, stretching down across the continent from Oregon to Appalachia. Temperate forests also covered Europe and Asia, a worldwide band of trees that botanists have named the Arcto-tertiary forest. In Europe, as we have seen, it was largely killed off during the Pleistocene Ice Age, but in Asia, where the mountain ranges also run north and south, more plants were able to survive the onslaught of the glaciers. Today the numerous genera that are found on both continents are survivors of that vast Arcto-tertiary forest.

In Asia, the woods have been cut repeatedly over the centuries and very little remains. In the cove forests of the Smokies, however, one can still walk among trees so venerable they easily recall that ancient time, and everywhere our eastern forest is filled with the young descendants of Arcto-tertiary trees. Of course, many species have been lost and many others added over millions of years, yet today, even after two centuries of human punishment, our woods are still a rich mosaic of plant associations, each shaped by exposure, soil, and drainage. In the next chapter we will examine some of them more closely.

II
THE COMMUNITIES

5
DECIDUOUS WOODS

Tulip Poplar
Liriodendron tulipifera

This discussion of deciduous woods has had to be oversimplified. Our eastern woods are an extremely complex plant community, consisting of hundreds of different species of trees, shrubs, and herbaceous plants in many different associations. The problem is further complicated by the fact that most woods are still in some phase of woody succession. Even so, they still contain a bigger variety of species of trees than are found anywhere else in the northern hemisphere.

The Mixed Mesophytic Forest

The tulip poplar, *Liriodendron tulipifera,* is an eastern tree of ancient ancestry and erect bearing. It is a member of the magnolia family. Flowers of the magnolia family are large and waxy with numerous styles surrounded by numerous stamens. Because they are so unspecialized, they are considered by some botanists to be early on the evolutionary scale. This is not to say, at least in the case of the tulip tree, that they are inefficient. Tulip trees are spry opportunists. They spring up in every available clearing and quickly rise to canopy level. They must to survive, because their seeds do not sprout in the dense shade of a mature forest. Young trees bloom when they are about fifteen or twenty years old, holding their flowers high in the sunshine, balanced like green and orange candle holders at the end of the branches. The seed pods look like wooden many-petaled flowers and stay on the branches most of the winter, making the tree easy to identify from afar.

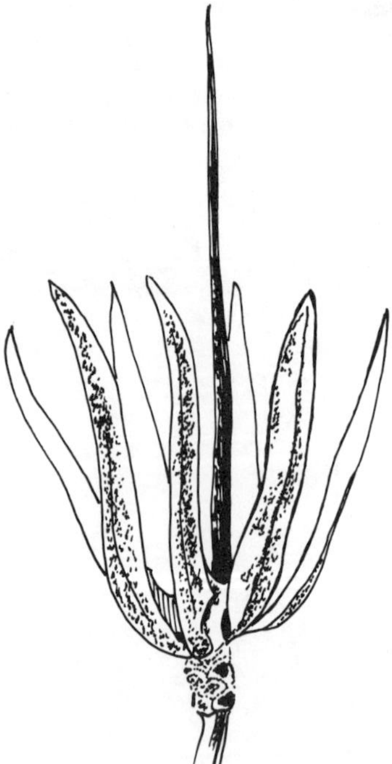

Seed Pod of Tulip Poplar
Liriodendron tulipifera

Tulip poplars are trees of deep, moist soils. They are conspicuous members of the mixed mesophytic forest, a botanical mouthful meaning a forest with no single dominant species growing in rich, well-drained soil. The richest mixed mesophytic forest grows in the southern Appalachians and is the evolutionary parent of all the other deciduous tree associations found in the eastern woods. It is the modern-day descendant of the great worldwide temperate forest of Tertiary times.

Trees endemic to the southern Appalachians that enliven the woods there as nowhere else include the silver bell, *Halesia monticola*. When it blooms in April, it sprinkles the treetops with white confetti, so numerous are its pristine bells. The Carolina silver bell, *H. carolina*, is a smaller species often grown in gardens. Both are hardy to eastern Massachusetts. The sweet buckeye, *Aesculus octandra,* too, is found only in these southern mountains, but it is a component of several forest types in a wide range of elevations. It resembles the horse chestnut, *A. hippocastanum,* an alien tree often planted in the Northeast. The sweet buckeye has the typical palmately compound leaves of the genus, the leaflets surrounding the center like spokes of a wheel. The greenish yellow flowers are held on showy, upright spikes during May and June.

Carlina Silverbell
Halesia carolina

Two deciduous magnolias also grow in these southern forests: the umbrella magnolia, *Magnolia tripetala,* a tree with giant, almost tropical-looking leaves, and the Fraser magnolia, *M. fraseri,* distinctive for its "ears" at the base of each leaf. Many southern plants have *fraseri* for a specific name. They are named for John Fraser, a Scottish collector of American plants who lived from 1750 to 1811 and kept a nursery in Chelsea, England. He made several expeditions to North America, collecting plants from Newfoundland to the Carolinas.

Other great trees of this forest include the sugar maple, *Acer saccharum,* the beech, *Fagus grandifolia,* and the hemlock, *Tsuga canadensis*, all familiar trees in New England. Also present are white ash, *Fraxinus americana*, and the black, white, and red oaks, *Quercus velutina, Q. alba,* and *Q. rubra,* which grow in mesic soils all over the east. The red maple, *Acer rubrum,* is here, too, with its deep carmine flowers in earliest spring and brilliant red foli-

Sweet Buckeye
Aesculus octandra

Fraser Magnolia
Magnolia fraseri

age in autumn. It is a tree found in nearly all our woods, apparently able to adapt to a variety of habitats. Here too is the black cherry, *Prunus serotina*, known to most of us as an opportunistic tree of thickets and hedgerows.

Along stream banks and bottom lands in the mixed mesophytic forest grow many trees that are also found in the low forests of the Coastal Plain and along river banks in the Piedmont: the black walnut, *Juglans nigra*, supplier of sweet nuts and superlative cabinet wood, and the mottled sycamore, *Platanus occidentalis*, as well as black willow, *Salix nigra*, and sweet gum, *Liquidambar styraciflua*. This last is a southern tree with star-shaped leaves as clean and crisp as a Christmas cookie, and dangling seed pods as prickly as a burr.

These are trees of the canopy, a botanist's name for the tallest trees in a forest community. Typically, a woods has four levels. Below the canopy are the understory trees, smaller species that tolerate the shade of the arboreal giants. Below them are the shrubs, and below them a ground cover of herbaceous plants and ferns. Lowest of all are the mosses. Each layer has adapted to progressively diminishing levels of light, so herbaceous plants, which need the least, are always the last to arrive in a developing woods.

The tallest tree of the understory is the black gum or tupelo, *Nyssa sylvatica*, although some individuals may grow large enough to reach canopy level. Its lustrous leaves catch and reflect the filtered sunlight and turn blood red in early autumn: the first in the forest to change. Tupelo is found throughout the east on dry, sandy soils as well as rich bottom lands. Botanists recognize a second variety, *N. sylvatica biflora*, which grows in heavy, wet soils on the Coastal Plain.

Dogwood
Cornus florida

Shadbush, *Amelanchier* spp., unfolds its lacy blooms in earliest spring, blooming along the banks of streams, often growing with two members of the birch family, Betulaceae, the American hornbeam, *Carpinus caroliniana*, and hop hornbeam, *Ostrya virginiana*. American hornbeam has bark like twisted licorice, while the bark of hop hornbeam peels off in cinnamon brown scales. In May the most conspicuous tree of the understory is the flowering dogwood, *Cornus florida*, an especially lovely sight when mixed, as it often is south of Maryland, with redbud, *Cercis canadensis*, in drifts of pink and white.

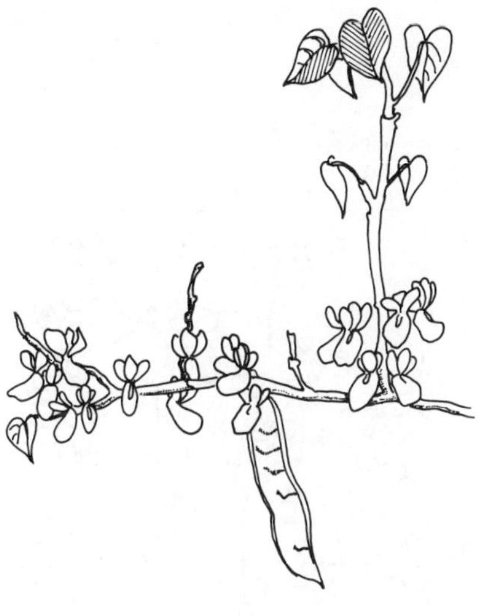

Redbud
Cercis canadensis

The shrub layer of the mixed mesophytic woods is a subtle parade of bloom beginning in early April when the spicebush, *Lindera benzoin*, flowers like a faint wash of yellow among trees still bare of foliage. By May, the mapleleaf viburnum, *Viburnum acerifolium*, blooms with pinxterbloom, *Rhododendron periclymenoides*, the exquisite wild azalea of the eastern woods. In June the white flowers of elderberry, *Sambucus canadensis*, billow like foam along stream banks, and in August the enticing fragrance of *Clethra acuminata*, the sweet pepperbush, perfumes the woods. The final curtain falls in mid-November when the improbable straplike flowers of witch hazel, *Hamamelis virginiana*, gleam among the falling leaves.

Vines are common in this forest as well, notably the Virginia creeper, *Parthenocissus quinquefolia*, and various species of wild grape, *Vitis* spp. Spiny greenbrier, *Smilax* spp., grows thick at the forest edge, and the great hairy ropes of poison ivy, *Rhus radicans*, climb the ancient trunks. Concerning poison ivy, the old saying, "Leaves of three, let it be!" is good advice for the amateur botanist. All parts of the poison ivy plant contain an oil that will cause an irritating rash in susceptible people. The plant takes many forms. It may be a high-climbing vine, a ground cover, or a thick, shrubby thicket two to three feet high. It does not twine like honeysuckle or bittersweet but clings to tree bark by means of holdfasts. In autumn, the shiny, three-part leaves turn a

Witch Hazel
Hamamelis virginiana

brilliant red, and clumps of dry, white berries cling to the plant well into winter.

Ferns are everywhere. The evergreen Christmas fern, *Polystichum acrostichoides,* is among the first to unroll its crosier in spring. Cinnamon fern, *Osmunda cinnamomea,* is easily identified by its fertile "cinnamon stick" that rises in the center, surrounded by sterile leaves, and New York fern, *Thelypteris noveboracensis,* has leaves that taper to a narrow point at both top and bottom. Interrupted fern, *Osmunda claytoniana,* may grow five feet high high in swampy areas. Its leaves are conspicuously "interrupted" by brown fertile pinnae in the middle of the stem. These are but a few of the more easily identified species. There are many more.

Hepatica
Hepatica americana

Woodland flowers begin with skunk cabbage, *Symplocarpus foetidus.* Their mottled purple hoods seem to appear overnight on moist bottom lands during the first warm days of March. Hepatica, *H. americana* or *H. acutiloba,* unfolds its delicate lavender blossoms soon after, often, like many evergreen species, on cool northern slopes where winter fluctuations of temperature are minimal. A rare flower of early spring in the Carolina mountains is oconee bells, *Shortia galacifolia,* which blooms among the laurel and rhododendron. Like Franklinia, shortia was once one of the lost plants of America. It was originally found in 1787 by the French botanist André Michaux, one of the most adventurous travelers and highly trained scientists of the eighteenth century. Michaux didn't name his find, but took a herbarium specimen home with him to Paris. Michaux died in 1803, and his collection languished unattended until the American botanist Asa Gray went to France in the 1830s. Intrigued by the little plant, he named it *Shortia galacifolia* after Charles Wilkins Short, a fellow botanist from Kentucky. At that time Gray knew only that it grew somewhere in North America. It wasn't until a simi-

Oconee Bells
Shortia galacifolia

lar Japanese species, S. *uniflora*, was found and described that he had a clue to its possible habitat. Even so, shortia was not found again in the wild until the 1880s, when Gray was an old man in the last years of his life. It is still one of our rarest plants.

Many fleeting flowers of the forest floor are called "spring ephemerals" because by early summer they have not only finished blooming, but have set seed and relapsed into dormancy. Plants need light in order to flower, and these species have evolved so as to take advantage of the warm spring sunlight at the only time it is available to them, before the leaves have developed. Trout lilies, *Erythronium americanum*, whose mottled leaves and nodding yellow flowers appear along stream banks in April, are spring ephemerals. So is the fragile spring beauty, *Claytonia virginica*, as well as Dutchman's breeches, *Dicentra cucullaria*, toothwort, *Dentaria* spp., and anemone, *Anemone quinquefolia*. All of these carpet the slopes of a mature forest. Scattered among them is the delicate rue anemone, *Anemonella thalictroides*, and the waxy three-petaled flowers of *Trillium* spp. Later, May apples, *Podophyllum pellatum*, jack-in-the-pulpit, *Arisaema atrorubens*, Solomon's seal, *Polygonatum biflorum*, and plume flower, *Smilacina racemosa*, appear. These are not spring ephemerals, as the plants remain all summer, setting fruit by late summer or fall. (Incidentally, jack-in-the-pulpit is only a "jack" while the plant is young. The tiny flowers that are contained within the hood, or spathe, start out as male and change to female as the plant ages. The plants that ultimately set fruit are "jills-in-the-pulpit"!)

These flowers fade as the tulip trees and magnolias bloom above them. By mid-June the spring ephemerals are gone and most tree growth is finished as well. The rest of the season is spent developing seeds and buds for the following year.

Trout Lily
Erythronium americanum

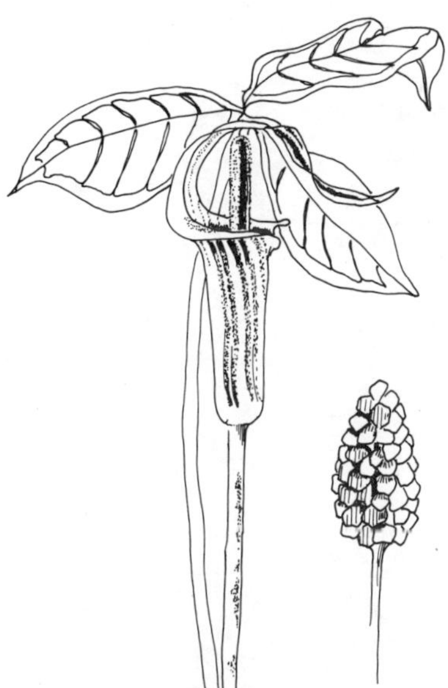

Jack-in-the-Pulpit
Arisaema atrorubens

Oak-Chestnut Forest

Among the evolutionary offspring of the mixed mesophytic forest is the oak-chestnut forest. This is the principal mesic forest of the Piedmont, that area of rolling countryside east of the Appalachians and west of the fall line, a term that originally referred to the waterfalls that marked the upper limit of navigation on the major rivers of the East. The name oak-chestnut is an obvious misnomer now, because the American chestnut, *Castanea dentata*, has disappeared from our woods. Virtually all the trees were killed by a fungus bark disease believed to have been introduced from Asia about 1900. From time to time sprouts still appear from remaining stumps, but the young trees rarely reach flowering size before succumbing to the blight. Botanists hope that a strain of the same fungus found in Europe that infects and inactivates the pathogenic strain will eventually make it possible to have chestnut forests once again. So far the method has been used successfully on individual trees, but the new strain has failed to spread widely on its own. So, for our time at least, the American chestnut is gone. Hickories have increased to fill the gap and some botanists now call this the eastern oak-hickory forest, to differentiate it from the oak-hickory forest that grows in dry soil on the western slopes of the Appalachians.

The difference between the eastern oak-hickory forest and the mixed mesophytic forest is principally in the number of species found in the canopy. As one travels north, many of the trees endemic to the southern mountains, such as the magnolias and silver bells, drop out of the woods' population. The tulip poplar is still abundant, however, and is quick to appear in a developing woodland or in a space left by a fallen tree. It quickly ascends to canopy height and persists in the mature forest as a large tree, often surrounded by smaller, later arriving oaks and hickories. White ash is also common in these woods, and red maple is everywhere, often growing in combination with beech on north-facing slopes, but the sugar maple is now found only on flood plains and bottom lands. Most of the understory trees, shrubs, and wildflowers of the mixed mesophytic woods are also abundant in this forest.

The oaks, central to many woodland communities, are a large and unruly clan. There are fifty-eight species of oak in North America, excluding Mexico, which has one hundred and fifty species of its own. In fact, our American oaks are thought to have migrated from Mexico thousands of years ago, spreading northward as the climate warmed. Southern oaks tend to be evergreen, a carry-over from their tropical heritage, and even deciduous northern oaks hang on to their leaves well into winter.

Oak trees have deep penetrating roots and the genus has adapted to a wide range of soil conditions. Chestnut oak and blackjack oak, *Quercus prinus* and *Q. marilandica*, are trees that grow on the driest soils of our region. Black oak, *Q. velutina*, likes moderately dry conditions, while white oak, *Q. alba*, and red oak, *Q. rubra*, are trees of the mesic forest. Willow oak, *Q. phellos*, water oak, *Q. nigra*, swamp white oak, *Q. bicolor*, and swamp chestnut oak, *Q. michauxii*, all grow in wooded swamps and wet bottom lands.

Generally, however, oaks are more tolerant of dry conditions than many trees, and there is evidence that where the land has been badly eroded or forests excessively cut, or where water tables have been depleted by overdevelopment, oaks tend to supplant other species, sometimes forming nearly pure stands. The area covered by oak forest has indeed increased since the early years of this century, but now they in turn are being seriously threatened by the rapid spread of the gypsy moth.

For a species that calls up the image of endurance, durability, and strength, oaks are notoriously slippery to identify. Botanical descriptions of oak trees are full of phrases such as "somewhat hairy," "somewhat thickened," or "usually but not always." To further confuse the amateur, leaves on young oak trees are very variable and often do not resemble those on mature specimens of the same species.

Oaks are members of the beech family, Fagaceae, an important group of timber trees as the wood is hard and durable and takes a good finish. Taxonomists have divided the oaks into two groups, the white and the red (also called the black). The white oak group have rounded lobes without bristle tips; their acorns mature in one year and are found on the current season's growth. The inner surface of the acorn shell (not the cup) is smooth, not hairy. Two familiar members of the white oak group are the white oak, *Q. alba*, a dominant member of the eastern oak-hickory forest, and the chestnut oak, *Q. prinus*, a tree of especially dry soils.

White Oak
Quercus alba

The trees of the red (or black) oak group have bristle tips on their leaves and acorns that take two years to mature so that both tiny new ones and larger old ones are frequently found on the tree in summer. The inner surface of the acorn shell is hairy. Members of the red oak group commonly found in the oak-hickory forest are *Q. rubra*, the red oak, and *Q. velutina*, the black oak, both large stately trees, and *Q. coccinea*, the scarlet oak, which is usually smaller. All three species apparently hybridize, producing many individuals with variants of the textbook descriptions. Common farther south is the Spanish oak, *Q. falcata*, also called the southern red oak. Mature Spanish oaks are large, handsome trees with distinctive leaves that are glossy on top but covered with orange fuzz beneath, making them conspicuous from afar, especially on a windy day.

Hickories are members of the walnut family. They too have deep taproots that enable them to adapt well to dry soils. Hickories usually appear in a developing woods after the oaks have begun to bear acorns, because the acorns attract squirrels which then bring in the hickory nuts. Prevalent among the hickories found in this community are the mockernut, *Carya tomentosa*, its common name apparently referring to the impossibly thick shell that surrounds a tiny amount of meat within, and the all but inedible bitternut, *C. cordiformis*. Less common is the shagbark hickory, *C. ovata*, whose long strips of peeling bark make it easy to identify. Shagbark has the sweetest nuts of all. They were once a staple food of the Indians, who ground and mashed them, mixing the paste with water to make a "hickory milk" that was used in cakes. A fourth widespread hickory is the pignut, *C. glabra*, the nuts of which were once commonly fed to swine.

Mockernut Hickory
Carya tomentosa

Where the soil is particularly dry, the character of the forest changes. This is often apparent on the tops of ridges and hills where the rock substrata is close to the surface and the topsoil thin, but it is also characteristic of soils that have been badly eroded from poor management in the past. Individual trees become smaller in the dry woods as the lack of moisture inhibits their growth, and there are many fewer species in the community.

Certain species of oak are always indicative of dry soils: the post oak, *Q. stellata*, the chestnut oak, the bear oak, *Q. ilicifolia*, and the more southern blackjack oak, *Q. marilandica*. They are sometimes mixed with dogwood and the adaptable red maple. New Jersey tea, *Ceanothus americanus*, is a conspicuous shrub of dry woods. So is *Rhus aromatica*, the fragrant sumac, a plant very different in appearance from the familiar sumacs of open fields. Particularly diagnostic of dry woods, however, is the presence of ericaceous shrubs, or plants belonging to the heath family, Ericaceae. One of these is sourwood, *Oxydendrum arboreum*, a particularly attractive small tree that is found on ridges south of Pennsylvania. Sourwood has white flowers very like those of the Japanese andromeda, *Pieris japonica*, so widely planted in suburban gardens. They form feathery sprays in July, conspicuous against the

Sourwood
Oxydendrum arboreum

Mountain Laurel
Kalmia latifolia

Wintergreen
Gaultheria procumbens

shiny foliage, and as an extra dividend, the seed pods continue to cling to the tree while the leaves turn coppery red in October.

Mountain laurel, *Kalmia latifolia,* is probably the most well-known member of the heath family. It is common along the ridges of the Shenandoah where it nestles among stunted oaks with trunks green with lichens and branches frequently broken by ice and wind. Mountain laurel has clusters of intricate pink and white blossoms, each flower like a tiny hoop petticoat. In each the stamens are tucked into pockets in the corolla, ready to be popped free by a pollinating bee. Mountain laurel also grows freely around stony ledges in New England as well as in the low oak and pine forest of the Coastal Plain.

Various species of blueberries and huckleberries, genera *Vaccinium* and *Gaylussacia,* are also members of the heath family that are plentiful in dry oak forests. They have flowers like dangling pink or white bells as well as edible fruit. *Gaultheria procumbens,* or wintergreen, is a low creeping heath that is also found in these woods. It has waxy, bell-like flowers in summer and leaves that yield the fragrant oil of wintergreen. The pealike red berry is fragrant, too, and edible besides.

Herbaceous plants of the dry oak woods might include snake-root, *Eupatorium rugosum,* wild columbine, *Aquilegia canadensis,* and sarsaparilla, *Aralia hispida.* Sarsaparilla was another plant in vogue in the eighteenth century, because it was thought to be a cure for venereal disease. At various times it has been made into a beer, smoked as a remedy for asthma, and used as a flavoring for soft drinks.

Oaks and hickories are also prevalent in the woods of southern New England, especially near the coast, but the principal community of interior New England is the hemlock, white pine, and hardwood forest.

Hemlock, White Pine, and Hardwood Forest

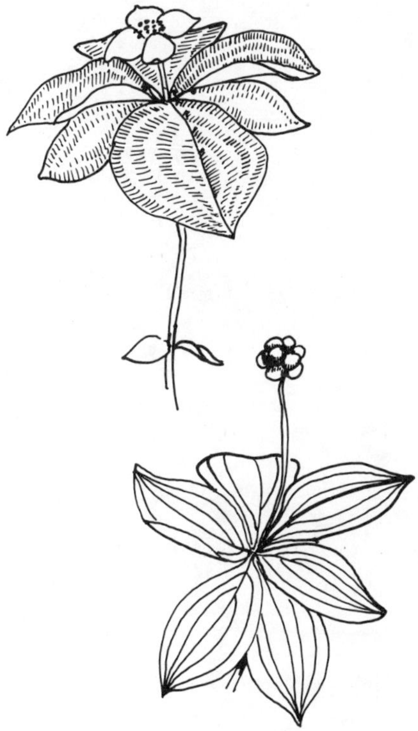

Bunchberry
Cornus canadensis

This is a forest of cool, moist climates. It is young in terms of evolution, because all of the area it now covers was once buried beneath the glaciers. Hemlock, *Tsuga canadensis*, and white pine, *Pinus strobus*, appear as patches of darkest green on the hillsides, mixed with hardwoods like sugar and red maple, yellow birch, *Betula lutea*, and white ash, *Fraxinus americana*. Red oaks and beech, *Fagus grandifolia*, are also common. The shrubs include many we have already mentioned, plus hobblebush, *Viburnum alnifolium*, yew, *Taxus canadensis*, and alternate leaved dogwood, *Cornus alternifolia*. Elderberry and witch hazel are also frequent. Ferns are plentiful and flowers include Canada mayflower, *Maianthemum canadense*, also called wild lily of the valley, besides many of the mesic varieties of the eastern oak-hickory woods. A particularly choice flower of northern New England is a tiny, ground-hugging dogwood called bunchberry, *Cornus canadensis*.

With the exception of coastal marshes and inland bogs, the New England forest was filled with ancient trees, many with trunks several feet in diameter, when the Pilgrims arrived in Plymouth in 1620. Wide boards taken from these trees may still be seen in early houses. Two hundred years later, the forest had been virtually eliminated. By 1850, only 5 percent of New England was still covered with trees, so today's woods are a little over a century old at the most. Many botanists think that pines and hemlocks were more common in the early days than they are today, principally because hardwoods will often sprout again from the roots. Conifers do not and are therefore forced to compete as seedling trees.

The Coastal Plain Forest

East of the Piedmont fall line is the Coastal Plain, a table-flat area of former sea bottom only a few feet above sea level. There is no Coastal Plain in New England; the rocky soil there was largely deposited by the glaciers and is quite different from the sandy sediment left by the sea. The boundary that marks the limit of the ice is evident as you travel south of New York City. The Coastal Plain spreads down the Atlantic coast from a narrow strip in northern New Jersey, continuing around the base of the ancient Appalachian continent into eastern Texas.

Because the Coastal Plain is prime agricultural land, few original woods remain. Those that do have been managed to such an extent that it is unclear what climax forest would develop, given the time and opportunity. Pines are common throughout. Pines, as gymnosperms, are of far more ancient origin than deciduous trees. Fossil records show that conifers much like those of today began appearing in the Triassic period, one hundred and eighty-five million years ago. Southern pines are quite unlike the glaucous, feathery white pine of the New England woods. They are stiff of habit and bristly of character. Some are quite contorted. The species most common on the Delmarva Peninsula, a representative area of Coastal Plain communities, are pitch pine, *Pinus rigida,* scrub pine, *P. virginiana,* and loblolly pine, *P. taeda.* Common further south are the shortleaf pine, *P. echinata,* and the longleaf pine, *P. australis.* The last is known colloquially as "broom pine" because its needles, which are frequently over a foot long, were once used to make brooms.

Loblolly Pine
Pinus taeda

Pines thrive on sandy soils that are less suited to deciduous forest and they are also found in conjunction with various species of the dry oak forest. We will examine this community more closely in the chapter on pine barrens.

Deciduous Woods

The deciduous woods of the Coastal Plain contain many trees found along stream banks and bottom lands within the mixed mesophytic forest, such as sycamore, *Platanus occidentalis*, black willow, *Salix nigra*, sweet gum, *Liquidambar styraciflua*, and black gum, *Nyssa sylvatica*, here the variety *biflora*.

The American holly, *Ilex opaca*, is common among the pines, particularly near the coast. So is the shrubby wax myrtle, *Myrica cerifera*, a southern evergreen cousin of the bayberry, *M. pensylvanica*. Both the sweet bay magnolia, *Magnolia virginiana*, and that stately tree of southern gardens, the bull bay, *M. grandiflora*, are trees of the Coastal Plain. Shrubs include both mountain laurel, *Kalmia latifolia*, here far from any mountains, and sheep laurel, *K. angustifolia*, also called lambkill for its toxic effect on sheep and other livestock. The sweet-smelling swamp and coast azaleas, *Rhododendron viscosum* and *R. atlanticum*, are special shrubs of the Coastal Plain. So is the winterberry, *Ilex verticillata*, noted for its brilliant berries at Christmas time. One distinctive small tree or shrub is Hercules'-club, *Aralia spinosa*, a relative of sarsaparilla, that frequently grows at the forest edge. It somewhat resembles a sumac, but has numerous prickly spines on its trunk and branches, and huge, creamy white clusters of flowers in August.

Sweet Bay Magnolia
Magnolia virginiana

We have already seen how higher altitudes favor northern or boreal plants. Similarly, the warm ocean current known as the Gulf Stream moderates the coastal climate so that many southern, or austral, plants reach the northernmost part of their range along the Coastal Plain. Loblolly pine is one of these. The word loblolly is a southern country word meaning mudhole or swamp. Pond pine, *P. serotina*, is another southern species that reaches its northern limit in the New Jersey Pine Barrens. So does water oak, *Q. nigra*, and its frequent companion, sweet gum. Blackjack oak, *Q. marilandica*, and willow oak, *Q. phellos,* are also southern species that extend north along the Coastal Plain. Persimmon, *Diospyros virginiana,* is another. Persimmon, a close relative of ebony, is a tree of primitive origin that dates from the early Cretaceous period, one hundred and thirty million years ago.

Woodland flowers are not as plentiful on the Coastal Plain, perhaps because truly mature forests are rare. However, numerous species of ground pines and club mosses, *Lycopodium* spp., carpet the forest floor.

6

OPEN LANDS

Forests can provide fuel and shelter, but agriculture must take place in the sun. So for thousands of years the development of civilization has resulted in the cutting of the forest. Forests and agricultural land are valued in densely populated Europe and have been meticulously managed for maximum fertility and economic production for centuries. Yet this was not always the case. In the ancient civilizations of the Mediterranean, the deforestation was so extensive and the land so badly abused that soils that were fertile in Classical times are now too eroded and poor to support the original vegetation. Unfortunately, Europeans coming to eastern North America found such a vast expanse of land that was both fertile and available that they quickly reverted to the earlier pattern of deforestation and abuse.

The woodland Indians of eastern North America were not an agricultural civilization. They lived primarily by hunting and fishing, although they did grow corn, beans, pumpkins, and squash in clearings near their villages. When the soil was exhausted, they simply abandoned the plot and cleared another, but as the Indian population was small and their villages widely scattered, there was little effect on the woods as a whole.

The European settlers, on the other hand, immediately set to work clearing land, putting up fences, and establishing farms, and within two generations, they had changed the whole character of the American landscape. The first land to be cleared was always mesic land: rich, moist, and well drained, for the land that grew the richest forest also grew the best crops. Early accounts by Europeans visiting this country invariably remark on the poor farming practices of the American farmer, complaining that they neither rotated their crops nor manured their fields. We know that crop rotation was not common in Pennsylvania until the 1780s, whereas by that time it had been practiced in England for over a century. But farm labor was scarce in the New World, and what there was was expensive, whereas land was there for the taking. There seemed to be no reason to husband land, so most American farmers simply cleared "fresh" land as it was needed. In Pennsylvania, the average eighteenth century farm was about one hundred and twenty-five acres. As much as thirty acres of this might lay fallow at any one time, often in what was called "bush fallow." A revealing term: it was probably exhausted land that had not been cultivated for several years.

Only about twenty acres of the typical Piedmont farm was left in woodland, usually on slopes too steep to plow. They were probably poor woods in any case, in terms of their plant life, for in those days it was common practice to let cattle and sheep graze in them. Woods became so scarce that as early as the 1790s Philadelphians were remarking on the lack of firewood, complaining that it had to be transported great distances to supply the city's needs.

About thirty acres of these early farms were kept as pasture or hayfields, while orchards and gardens were planted near the homesteads. The colonists brought over the seeds of forage grasses, fruit trees, vegetables, and medicinal herbs as well as many garden flowers. Many of these plants had become adapted to disturbed soil conditions over thousands of years of civilization. They flourished in the newly opened soil of the New World, especially as there was little competition from native plants adapted to a forest habitat, and soon escaped cultivation.

It is surprising how many common field flowers of today date from these early times. For instance, flax, *Linum usitatissimum,* was an important field crop in early Pennsylvania, grown for its fiber and oil. It escaped to grow along modern roadsides. Queen Anne's lace, *Daucus carota,* is really a carrot. It and dandelion, *Taraxacum officinale,* were grown as vegetables in New England before 1700. Indeed, dandelions have been an important food plant for centuries, and directions for growing them were given in several early herbals. In 1871, four varieties of dandelion were exhibited at the Massachusetts Horticultural Society, and by 1884 a Mr. Corey of Brookline was growing them from seed for the Boston market. Nearly every part of the plant is edible, and the leaves have more vitamin A than most greens that are commonly cultivated today. However, they tend to be bitter unless picked very young, and must be cooked in two or three changes of water.

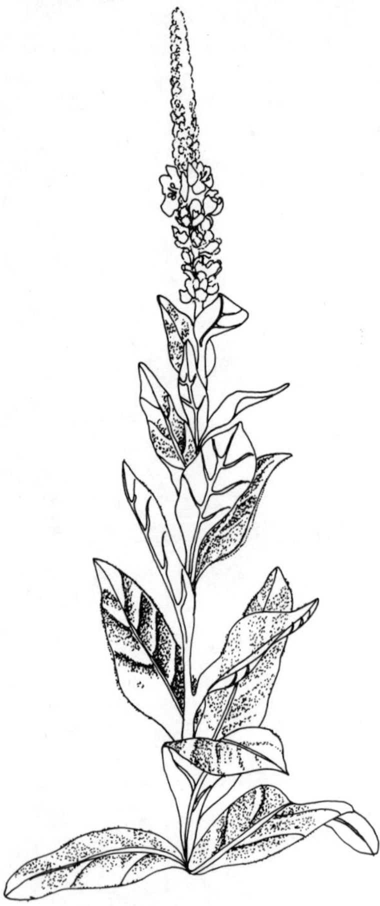

Flannel Mullein
Verbascum thapsus

Tansy, *Tanacetum vulgare*, yarrow, *Achillea millefolium*, and feverfew, *Chrysanthemum parthenium*, were grown as medicinal herbs. Flannel mullein, *Verbascum thapsus*, was used for everything from curing warts to dyeing hair, while teasel, *Dipsacus sylvestris*, was grown for its prickly seed head, which was used to card wool. Butter and eggs, *Linaria vulgaris*, common buttercups, *Ranunculus acris*, and field daisies, *Chrysanthemum leucanthemum*, are all on early garden lists, apparently brought by homesick colonists to remind them of home. The knapweeds, *Centaurea* spp., were also brought as garden flowers. We have only one American species of knapweed, *C. americana*, and it is a native of the Southwest. Celandine, *Chelidonium majus*, and bouncing bet, *Saponaria officinalis*, are also on the early lists. *Hemerocallis fulva*, the tawny day lily that brightens our roadsides in July, was nicknamed the "outhouse flower" by midwestern settlers because it grew so luxuriously in that location. How it has spread so far is a botanical mystery as it has been identified as a triploid form called "Europa" that never sets seed! Apparently the smallest root segment will suffice for a new plant to develop.

Most of the invasive prickly thistles of our meadows and fields are also aliens, possibly arriving as stowaways in shipments of grain. The Canada thistle, *Cirsium arvense*, which is not Canadian but European, is the most aggressive of the lot. It is a perennial that spreads by underground stolons as well as by far-flying thistledown, and will quickly dominate a pasture or field. It was such a pest of newly cleared land that Vermont passed the first antithistle law in 1795, requiring farmers to eliminate it. Many states still have such laws on their books. Bull thistle, *Cirsium vulgare,* and nodding thistle, *Carduus nutans,* can also be serious pasture pests because livestock crop down the competing forage, allowing them to flourish unchecked. They are biennials, however, and do not have invasive root systems, so mowing a field before the flower goes to seed will control them. That thistles are an ancient pest is evident from Genesis: when Adam and Eve were banished from Paradise, God cursed the ground so it would grow only thistles.

We do have several native thistles, all of which are more choosy in their habitats. Three of them might be found in fields and pastures, all belonging to the genus *Cirsium,* which differs from genus *Carduus* by the tiniest botanical technicality having to do with the hairs on the pappus, or modified calyx. Pasture thistle, *C. pumilum,* is our largest flowered species with heads two or three inches in diameter. Field thistle, *C. discolor,* is distinguished by a felt of white wool on the undersides of the leaves, while tall thistle, *C. altissimum,* lives up to its name by growing up to twelve feet tall.

The new settlers also grew in their gardens native American flowers, many of which later traveled back to England to be hybridized as garden flowers. Thus, New England and New York asters became England's Michaelmas daisies. The colonists also grew sunflowers and wild geraniums as well as hepatica, blood-root, lilies, lobelias, iris, and violets.

Plant communities of open fields and meadows are dominated by grasses, legumes, and flowers of

Field Thistle
Cirsium discolor

the daisy family, Compositae, which are also called Asteraceae.

Grass

It is odd that people pay so little attention to grass. Not only is the grass family, Gramineae, the largest family of flowering plants worldwide, but it is also the most useful to man. Grasses include all the grain crops: wheat, barley, rice, oats, and rye, as well as Indian corn, the only grain native to North America. Grasses mixed with legumes make up the forage that feeds our livestock. Grass stems have been used to thatch roofs, weave baskets, and bed animals. Lacking trees, the early Nebraska pioneers even built houses from grass sod. In nature, grasses are important because they are the first plants to clothe barren land, preventing wind and water erosion.

Grasses are monocots, a subdivision of flowering plants. When the seed of a monocot sprouts, it has only one seed leaf. Monocots, which include, among others, the lily and iris families, also have straplike leaves and flower parts in multiples of three.

Grass flowers are wind-pollinated, and except for the heavier grains, their seeds are small and easily dispersed. The entire flowering part of a grass plant is called the ear, while each individual flower is call a spikelet. A spikelet consists of two husks that contain the stamens, pistil, and eventually the fruit. Grass stems, called culms, are usually hollow, but they have solid joints, called nodes. The leaves, which are attached at the nodes, consist of a sheath enclosing the culm and a blade which is long and narrow. Grasses have persistent, adaptable roots that spread out in damp weather and bore downwards in dry. The roots of native prairie grasses can store enough food to outlast years of drought. The leaves and sheaths of all grasses readily regrow when cut, making grass a perpetual grocery for grazing animals.

Imported Forage Grasses

Man and his herds have followed the grass for ten thousand years, so it is not surprising that European grains and forage grasses followed the colonists onto their farms in the New World. A common pasture grass, still prevalent on open land all over the East, is Kentucky bluegrass, *Poa pratensis,* actually a native of Europe and Asia. It was first imported to England in the eighteenth century. The American colonists called it "English grass" until about 1720, but as an important range cattle business began to develop in the lower Piedmont states, "English grass" became known as Kentucky bluegrass. *Poa pratensis* now appears spontaneously throughout the East where the humid climate suits it well. Because it forms a dense sod, it is popular for lawns, but this same quality in open meadows tends to inhibit field flowers from seeding.

While successful pasture grasses are able to withstand both trampling and continual cropping, certain other plants thrive in pastures simply because they are unpalatable to livestock. These include the thistles we have already mentioned, but also buttercups, *Ranunculus acris*, horse nettle, *Solanum carolinense,* a virulently poisonous member of the tomato family, and poke, *Phytolacca americana.* Young sprouts of poke are sometimes eaten like asparagus, but great care must be taken because other parts of the plant are poisonous to humans. Experiments have shown that the seeds of poke must travel through the digestive systems of birds or animals before they will germinate. They may then lie dormant in the ground for years waiting for the opportunity to grow. Poke will quickly colonize disturbed soil and often sprouts along paths worn thin by cattle or horses. The berries were once used for dye.

Timothy, *Phleum pratense,* is a grass which when planted in combination with alfalfa makes superior horse hay. There are various stories about its

Horse Nettle
Solanum carolinense

arrival here. According to one, it was imported by a New Hampshire farmer named Timothy Hansen in 1731 after he took a trip to France in search of rich grass for his pastures. Somewhat immodestly, he named the new grass for himself and immediately began marketing the seeds. Incidentally, the specific name *pratense* or *pratensis* means "of meadows." Not surprisingly, it is attached to many forage plants.

Orchard grass, *Dactylis glomerata*, also came by way of England, where it was called cocksfoot because of its typical three-part ear. Orchard grass grows well in light shade and acquired its American name because farmers planted it in their orchards.

In recent years, highway departments have been responsible for the spread of certain grasses. Fescues, especially *Festuca elatior*, the meadow fescue, were planted along many roads because they grow well in nitrogen-poor soil, while the love grasses, *Eragrostis* spp., were imported from the dry plains of Africa for their drought resistance. Bermuda grass, *Cynodon dactylon*, an alien that is as tough and persistent as crab grass, was also used along roads.

Native Grasses

Because mesic land in the East was originally covered with woods and grass does not grow in dense shade, there are relatively few indigenous species. Two conspicuous eastern grasses, however, are broomsedge, *Andropogon virginicus*, which often spreads into exhausted fields, and purple top, *Triodia flava*, a waxy, panicled grass noted for its attractive purple ear in late summer and fall. Many *Panicum* species, called switch grasses, are also attractive natives as are all the useful grasses of marsh and beach that we will examine in a later chapter.

Native grasses are warm season grasses, starting into growth after the soil warms in late spring and spurting upward during the heat of summer. The European forage grasses are cool season grasses. They green up quickly in early spring and flower in early summer, then fall dormant during the hot days of July and August. When the cooler days of autumn arrive, they begin again to grow and often stay green all winter.

One cannot talk of native American grass without mentioning the prairie. The first settlers to move west found a land beautifully described by Willa Cather in *My Antonia*:

> As I looked about me I felt that the grass was the country, as the water is the sea. The red of the grass made all the great prairie the colour of wine-stains, or of certain seaweeds when they are first washed up. And there was so much motion in it; the whole country seemed, somehow, to be running.

The blowing grass was turkey foot, or big bluestem, *Andropogon gerardi*; it was the backbone of the tall grass prairie. In moist years, big bluestem may grow six feet high. In Willa Cather's day a single stand might have been thousands of years old, the tops growing fresh each year from ancient, enduring roots that reached many feet into the earth. The name bluestem comes from a faint blue-green cast to the leaves; the stem is actually tinged with purple. The flowering ear is a three- to five-fingered raceme, resembling a turkey's foot.

The grass Willa Cather describes as "red as wine stains" was probably little bluestem, *Andropogon scoparius*. It grew where the prairie was higher and drier and was often mixed with leadplant, *Amorpha canescens*, and purple prairie clover, *Petalostemum purpureum*, two midwestern members of the pea family.

The third great grass of the tall grass prairie was Indian grass, *Sorghastrum nutans*, a grass that has migrated east in many places. It, too, matures at six feet by the end of summer and has graceful wandlike ears. The foliage turns an attractive yellow orange in autumn. Switch grass, *Panicum virgatum,* is a prairie grass now common in the East, easily recognized by its delicate lacy ears that are held above a dense clump of straplike foliage.

Evening Primrose
Oenothera biennis

The grass that the early pioneers used to build their sod houses was buffalo grass, *Buchloe dactyloides.* It is the dominant grass of the western Great Plains, where the soil is drier than that of the tall grass prairie. Buffalo grass seldom grows higher than five or six inches and forms a tangled mat that resists invading plants.

Prairie flowers that have spread eastward in large numbers include various species of sunflower, *Helianthus,* as well as black-eyed Susans, *Rudbeckia hirta,* and butterfly weed, *Asclepias tuberosa.* Many others belong to genera that also have eastern species, such as ironweed, *Vernonia,* wild indigo, *Baptisia,* golden ragwort, *Senecio,* and evening primrose, *Oenothera,* as well as the more familiar, *Phlox,* gentians, *Gentiana, Aster,* and the ever-present goldenrod, *Solidago.*

The prairies are gone now and many of the fields and pastures of the East have vanished under housing developments and shopping malls. Many others are well on their way into woody succession. Paradoxically, roadsides, where they are not continually mowed or sprayed, are becoming the last refuge of meadow flowers. In many parts of the country this trend is being actively encouraged in a program called "Operation Wildflower," run by the National Council of State Garden Clubs, Inc. In the Midwest, some groups are trying to restore the prairie ecosystem on plowed fields and worn-out pastures. This is a complicated task, for prairie grasses and wildflowers are not disturbed soil plants but part of an ecosystem that took thousands of years to evolve.

Legumes

Legumes are members of the pea family, Leguminosae, second only to grasses in providing food for man and livestock. A species of bacteria able to convert the nitrogen gas in the atmosphere into compounds usable by green plants lives symbiotically on the roots of leguminous plants such as beans, clover, alfalfa, and peas. Because they are thus able to thrive in nitrogen-poor soils while enriching them in the process, they are invaluable to farmers, who grow them as livestock feed and use them as "green manure." Legumes include all the commercially grown beans and peas, including soybeans, a plant with so many uses that it was traditionally referred to in Japan as "the little honorable plant." Alfalfa, *Medicago sativa*, red clover, *Trifolium pratense*, and alsike clover, *T. hybridum*, are all legumes that are widely cultivated as forage plants. Alfalfa has the most ancient lineage. It is alluded to in Babylonian texts of 700 B.C., and Pliny recorded that the Persians who invaded Greece in 490 B.C. brought alfalfa to feed their chariot horses.

Red Clover
Trifolium pratense

White clover, *T. reptens*, used to be a common component of lawns, but it has been largely eliminated by the use of broad-leaf weed killers. Unfortunately, its nitrogen-fixing ability has disappeared as well. Other alien but widespread clovers found growing wild in our fields include black medick, *Medicago lupulina*, yellow sweet clover, *Melilotus officinalis*, and bird's foot trefoil, *Lotus corniculatus*. Crown vetch, *Coronilla varia*, is a legume that is widely used along highways.

Kudzu
Pueraria lobata

Kudzu, *Pueraria lobata*, is an infamous member of the leguminous tribe. During the dust bowl days of the thirties, kudzu was hailed as a wonder plant that would rejuvenate soil and stabilize erosion on worn-out cotton land in the South, while at the same time providing cheap, fast-growing, and nutritious forage for livestock. Not only that, glowed the promotional literature, it grew up to a foot a day and had the amazing ability to produce a new plant from every node along the stem! Thousands of workers in the Civilian Conservation Corps went south to plant kudzu along highways. Farmers stopped growing cotton and grew kudzu instead. By the end of the Second World War, small southern towns were holding kudzu festivals, a kudzu queen was crowned in Alabama, and Japan and China were both claiming to be the native home of this wonderful plant. Alas, it wasn't long before kudzu was invading valuable cropland and smothering acres of timber with an impervious curtain of green. Now the problem is to get rid of kudzu, which squats on the southern landscape like an inscrutable symbol of human ignorance.

Less flamboyant native legumes include ground nuts, *Apios americana,* wild bean, *Phaseolus polystachios,* milk pea, *Galactia regularis* and hog peanut, *Amphicarpa bracteata.* All, as their common names suggest, were once used for food by the Indians and early colonists. Various vetches, vetchlings, and lespedezas are native legumes, as are the numerous tick trefoils, *Desmodium* spp., that are rarely noticed until the seed pods, known as beggar's lice, cling to one's clothes in the fall.

Flowers of the pea family are bilaterally symmetrical. They have five petals, but the two lower ones are often joined to form a "keel" that surrounds the stamen and pistil. The two side petals form wings, and the upper one a "banner." Sometimes, as in the clovers, the individual flowers are minute, but are crowded together in a head.

Flowers of Mesic Fields

Meadow flowers are dominated by members of the Compositae, or aster family. Composites are the most advanced family of dicots. Dicots, as opposed to monocots, have two seed leaves, flower parts in multiples of two or five, fibrous roots, and net-veined leaves. Each composite flower is really composed of many flowers. A daisy, for instance, is made up of tiny disk flowers, each with its own stamens and pistil, surrounded by ray flowers, again each with stamens and pistil. The tiny individual blossoms can be readily seen with the help of a hand lens. Some composites lack ray flowers, being composed entirely of disk flowers, but all have a structure beneath the flower that is composed of tiny leaflets, or bracts. This is called an involucre.

Among the first natives to bloom is fleabane, *Erigeron annuus,* a prairie flower that spread eastward as the forest was cleared. It has small daisylike flowers with narrow rays and looks rather like a white aster, except that it begins blooming in May and continues for most of the summer. Golden ragworth, *Senecio aureus,* soon follows, another daisylike flower with butter yellow rays and disks. "Nature's first green is gold," said the poet Robert Frost, and golden ragwort is followed by drops of yellow dandelions, orange hawkweeds, *Hieracium* spp., and swales of yellow mustards, *Brassica* spp. At their feet may be creeping yellow moneywort, *Lysimachia nummularia,* hiding among the violets in the developing grass. We have dozens of native violets. Most are woodland plants, but the common blue violet, *Viola papilionacea,* is found in open meadows, and careful observation may reveal other species as well. Fragile Quaker ladies, *Houstonia caerulea,* may spread over a grassy bank like tiny lavender dots, while Venus' looking glass, *Specularia perfoliata,* hides among the grasses, too. Its name comes not from the stalkless sky blue flowers but from the flat, polished seeds that follow them.

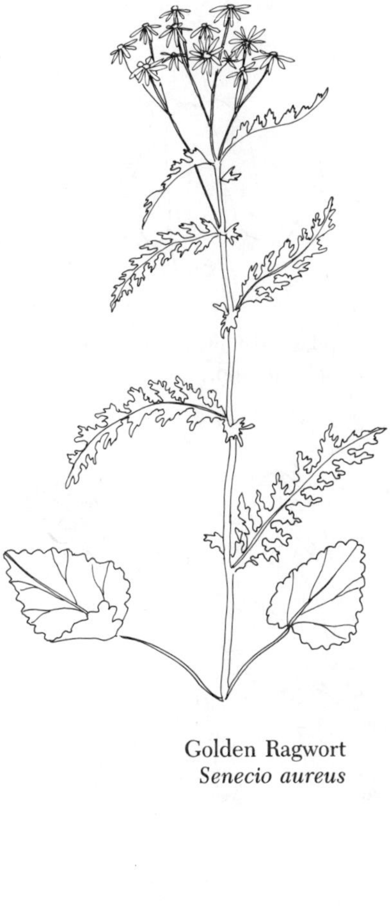

Golden Ragwort
Senecio aureus

The first yarrows, *Achillea millefolium,* bloom in June. The flowers of some are not the usual creamy white but a subtle dusky pink. By late June the common milkweed, *Asclepias syriaca,* unfolds heavy

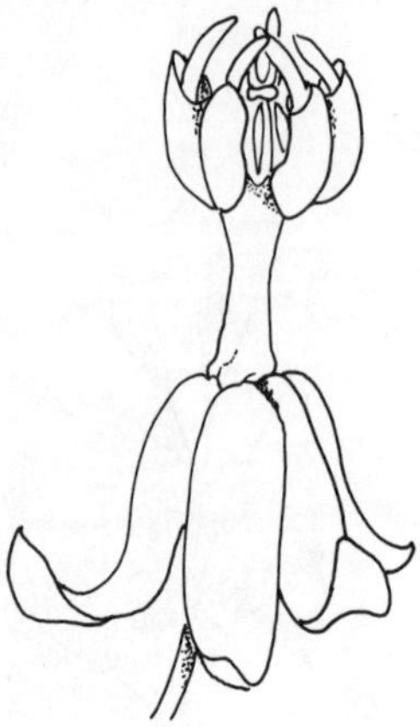

Milkweed Flower (enlarged)
Asclepias spp.

Turk's-Cap Lily
Lilium superbum

pink blossoms that hang like fragrant popcorn balls in the leaf axils. Milkweeds have distinctive, intricate flowers. The five petals are attached to a club-like column and hang down rather like a tiny hula skirt, hiding the sepals. The stamens and pistil are concealed behind "hoods" at the top of the column and each hood ends in a peculiar hornlike structure. The purpose of this strange arrangement is to discourage any insects but bees and butterflies from pollinating the flowers. Only their strong legs can bend down the hoods and reach the pollen.

If you are lucky, you may find a pasture rose, *Rosa carolina*, blooming in your meadow. It is a shy rose with single shell pink flowers, nowhere near as aggressive as the Asian invader, *Rosa multiflora*. Soon, the slender blue-eyed grasses, *Sisyrinchium* spp., dot the true grass with tiny blue six-point stars. They are members of the iris family, Iridaceae. Looking at first glance like a pink species is the alien Deptford pink, *Dianthus armeria*, a diminutive member of the carnation family, Caryophyllaceae.

By July first, the meadow belongs to the robust black-eyed Susans, *Rudbeckia hirta*, a prairie flower that has moved east in abundance. Mixed with St. John's wort, *Hypericum* spp., they stain the summer meadow gold accented here and there with the pale yellow of the common evening primrose, *Oenothera biennis*. Butter and eggs, *Linaria vulgaris*, blooms like tiny orange and yellow snapdragons, while chicory, *Cichorium intybus*, and Queen Anne's lace, *Daucus carota*, decorate the roadsides with blue satin and eyelet.

By August early goldenrod appears, *Solidago juncea*, and all the weedy plants with homely names reminiscent of vanished barnyards: lambsquarters, *Chenopodium album*, goosefoot, *C. hybridum*, and sow thistle, *Sonchus* spp. The air is filled with the hot, dusty smell of ripening corn and Turk's-cap lilies, *Lilium superbum*, nod along the stone walls of New England like opulent visitors from the Orient.

Each exotic recurved flower has a green star in its throat and conspicuous stamens like small brown polo mallets. Lilies seem far too tropical for these northern climes, but in fact, some twenty species grow wild in North America. All are stately plants of regal bearing, fit for the finest garden.

In early September the goldenrods and asters color the landscape yellow and mauve. There are seventy-five species of goldenrod native to the East and they vary considerably in height, habit, and choice of habitat. The two most common in old fields and pastures are the tall goldenrod, *Solidago altissima,* and the rough-stemmed goldenrod, *S. rugosa.* They are easily distinguished by their leaves. Tall goldenrod has narrow leaves with three parallel veins, while the leaves of rough-stemmed goldenrod are feather veined. Both have stout, hairy stems, although tall goldenrod's is more greyish and downy. A third common species is Canada goldenrod, *S. canadensis,* easily identified because it is the only species attacked by a midge larvae that causes crown gall, a conspicuous proliferation of leaves at the apex of the stem. The spherical swelling farther down the stem found on several species is caused by the larva of a fly. The fly lays its eggs on the stem. When the larva hatch and begin to feed on the plant's tissue, it reacts by encasing them in a gall.

New England asters, *Aster novae-angliae,* bloom on moist or mesic fields, forming stout clumps of royal purple daisies as beautiful as any garden perennial. In fact, most garden asters are selections of this species; nurserymen collect seedlings that vary from the standard color and then propagate them vegetatively. The last plants to bloom before winter are the white asters or frostweeds, *Aster ericoides, A. pilosus, A. simplex,* and others. They are stout, bushy plants with myriads of tiny white daisies, called frostweeds because they go on blooming long after everything else has finished. The yellow centers to the flowers usually turn red-

New England Aster
Aster novae-angliae

dish brown after the plant is fertilized, but in at least one species, *A. lateriflorus*, the calico aster, they turn a distinctive purple.

Flowers of Dry (Xeric) Fields

Dry meadow plants are found on old fields that have lost topsoil to erosion, along the shoulders of graded roads or on sandy soil near the coast. Imported forage grasses do not grow well on dry soils, so on dry hillsides they are quickly replaced by broomsedge, *Andropogon virginicus*. In especially dry places, poverty grass, *Aristida dichotoma* is common as well.

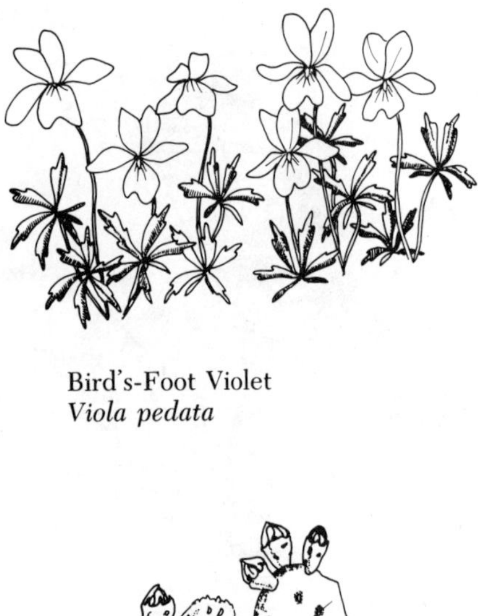

Bird's-Foot Violet
Viola pedata

Our most beautiful native violet, the bird's-foot violet, *Viola pedata*, blooms in May on sterile, sandy soil. It has large, almost pansylike flowers of a vivid blue-violet and finely cut leaves rather like those of an anemone. Bird's-foot violets take up to three years to bloom from seed, although once established they may persist for a decade. Prickly pear cactus, *Opuntia humifusa*, is the only true cactus that is widespread in the East. It has jointed flat pads, which are actually modified stems, and tufts of bristles. Prickly pear was once imported to Australia as a living fence to segregate cattle and went wild over the landscape in the same way kudzu has here. Other dry meadow forbs include the pussytoes of the genus *Antennaria*, which unfold their silvery "paws" in June. Their cousins the everlastings come on in August and September. The most showy of these is the perennial pearly everlasting, *Anaphalis margaritacea*, which may form large colonies in sandy soil. The annual cudweed, *Gnaphalium obtusifolium*, is less showy but has a delicious spicy scent of curry.

Prickly Pear Cactus
Opuntia humifusa

Eupatorium hyssopifolium is a dry meadow species of boneset and *Solidago nemoralis* a dry-loving goldenrod. It is a compact, almost dwarf plant with handsome gray foliage. *Monarda fistulosa,* the lavender bergamot, and *Liatris* spp., the blazing stars, are prairie plants that have spread east to dry meadows in some places. Butterfly weed, *Asclepias tuberosa,* the handsome orange milkweed of dry hillsides and old railroad embankments, and *Coreopsis lanceolata,* the yellow coreopsis, are also flowers of the prairie.

Goatsbeard, *Tragopogon pratensis,* looks like a huge flat-topped dandelion with square petals and has a seedhead like a three-inch pom pom. Ground cherry, *Physalis* spp., is a dry-loving member of the tomato family, Solanaceae, that bears sweet yellow fruit concealed within a papery husk. But beware you don't confuse it with horse nettle, *Solanum carolinense,* another member of the same family, which is deadly poisonous! Oddly enough, horse nettle was once given in minute doses as a cure for tetanus.

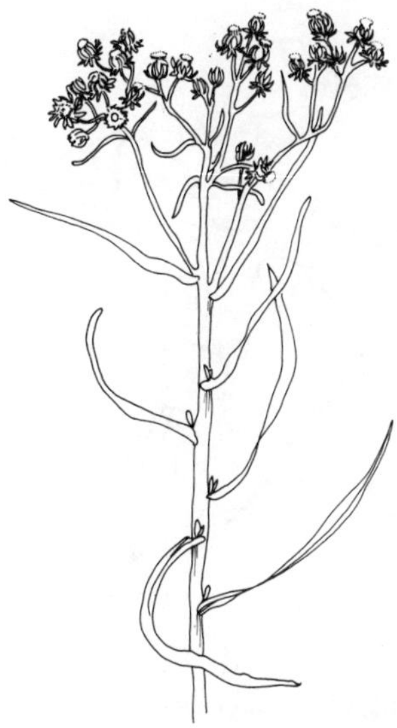

Pearly Everlasting
Anaphalis margaritacea

Flowering spurge, *Euphorbia corollata,* is another prairie plant that grows in the east on dry sandy soils. It is a smooth, almost succulent plant with showy white flowers rising from a whorl of smooth green leaves. What appears to be the petals of the flower are really bracts that surround the tiny true flowers in the center. It blooms in August. Two of our most beautiful asters bloom on sandy soils: *A. spectabilis,* the showy aster, and the stiff aster, *A. linariifolius,* a dwarf plant with leaves as stiff as spruce needles. Both have lovely lavender flowers.

Flowering Spurge
Euphorbia corollata

Flowers of Wet (Hydric) Meadows

Wet meadow species congregate along the course of a stream or in a poorly drained swale or roadside ditch. They may also be clustered at the bottom of a hill, where the soil is rich and moist, perhaps even in the same field as some of the upland species, although the two groups never intermix.

Virginia Bluebells
or Brandywine Bluebells
Mertensia virginica

Green-Headed Coneflower
Rudbeckia laciniata

In April and May, the golden alexander, *Zizia aurea*, blooms along brooks and in low meadows. It looks rather like a yellow Queen Anne's lace, and is in fact a member of the same family, Umbelliferae, the parsley family. Two other yellow umbelled flowers that may be confused with it are the meadow parsnip, *Thaspium trifoliatum*, and the alien wild parsnip, *Pastinaca sativa*. The central cluster of flowers on *Zizia* are stalkless while those of *Thaspium* are stalked, a distinction easily made with a hand lens. *Pastinaca* is never found in wet soils. It is a stout upland species that spreads over dry hillsides.

Two native white mustards, family Cruciferae, bloom in damp ground in early spring. They are the Pennsylvania bittercress, *Cardamine pensylvanica*, and spring cress, *C. bulbosa*. But perhaps the most beautiful flower of streamsides and flood plains is the Virginia bluebell, *Mertensia virginica*. Its clusters of nodding pink buds open to pure blue bell-like flowers in May.

In June the fringed loosestrife, *Lysimachia ciliata*, blooms. Its single yellow flowers, each with tattered petals, nod from the leaf axils. Nearby, elderberries, *Sambucus canadensis*, and the gray-stemmed dogwood, *Cornus amomum*, may bloom with a swamp rose, *Rosa palustris*. Swamp rose is larger than the upland pasture rose and has flowers of a deeper pink. Its thorns form tiny hooks, while those of the pasture rose are straight.

As June moves into July, the green-headed coneflowers, *Rudbeckia laciniata*, appear on flood plains and river banks, while spotted jewelweed, *Impatiens capensis*, and its yellow sister species, *I. pallida*, begin to show color along every brook and watercourse. The sepals of each jewelweed flower form a spur full of sweet nectar that attracts bumblebees, hawkmoths, and hummingbirds. Jewelweeds are annuals that have evolved many tricks to insure the healthy survival of the species. For instance, each flower produces ripe pollen before it exposes a stigma, thus neatly avoiding self-fertilization that might gradually weaken the species. However, in the event of a year so dry it limits flowering altogether, the plants produce special self-fertile flowers, as a sort of emergency measure to insure seed production. But the jewelweed's system of seed dispersal is perhaps the most spectacular of all its talents. The merest touch of a ripe pod causes it to spring open with a snap, flinging the seed in all directions. No wonder children call them "touch-me-nots."

Touch-Me-Not
or Jewelweed
Impatiens capensis

One of our most beautiful native lilies blooms on open, moist ground in July. This is the meadow, or Canada lily, *Lilium canadense*. It grows over six feet tall, dangling waxy, bell-shaped flowers from long stems that rise from a whorl of smooth leaves. July is also the month that the dusky pink spires of hardhack, *Spiraea tomentosa*, appear, often mingled with swamp milkweed, *Asclepias incarnata*. In this species the typical milkweed flowers are collected in a flat-topped inflorescence of iridescent pink.

By August, the Joe Pye weeds begin; their domes of pink froth rise on purple-spotted stems above a whorl of rough textured leaves. Our largest, most stately Joe Pye is the hollow-stemmed *Eupatorium fistulosum*, which towers a sturdy six feet or more. Joe Pye is believed to have been the name of an Indian herbalist who lived near Salem, Massachusetts and used this plant as a cure for fevers.

As summer begins to wane, the spires of cardinal flower, *Lobelia cardinalis*, glow like red velvet in low meadows and pondsides. Its sister species, the blue lobelia, *Lobelia siphilitica*, blooms among the creamy bonesets and yellow lance-leaved goldenrod, *Solidago graminifolia*. The bold New York ironweed, *Vernonia noveboracensis*, holds up shaggy flowers of a deep red-violet. The color is a perfect

Jerusalem Artichoke
Helianthus tuberosus

Sneezeweed
Helenium autumnale

foil for the pale lavender of *Aster puniceus*, the purple-stemmed aster, the most abundant species along streams and in low meadows of the Piedmont. Mixed among it one might find monkey flower, *Mimulus ringens*, or turtlehead, *Chelone glabra*. If you squeeze the flower, the "turtle" will open its mouth.

August and September is also the time for the sunflowers, *Helianthus* spp., their cousins the sneezeweeds, *Helenium* spp., and the tickseed sunflowers, *Bidens* spp. When most people think of sunflowers, they think of *Helianthus annuus*, the common sunflower that is the parent of all the large-disked giants grown in gardens. Indians used the seeds of this species for both food and oil, and this is thought to be the reason it is so widespread in the Midwest. Now it has come east as well and oddly enough grows abundantly near the Philadelphia Airport, raising its circular yellow heads from every drainage ditch, usually out of a bed of beer cans, and often in concert with the brilliant spires of purple loosestrife, *Lythrum salicaria*, a showy but invasive immigrant from Europe. A more wide-ranging sunflower in the East is the Jerusalem artichoke, *H. tuberosus*. The name Jerusalem artichoke supposedly comes from the Italian word for sunflower, "girasole." It is no relation of the artichoke, but its root is an edible rhizome that tastes rather like water chestnuts. It is sometimes sold in supermarkets under the name of "sunchoke."

Tall sunflower is another species often found in low meadows in the East. It is aptly named *H. giganteus* for it sometimes grows to ten feet or more, towering over the surrounding vegetation. Tall sunflower has stout red stems and rough, lance-shaped leaves.

Sneezeweeds, *Helenium* spp., differ from the true sunflowers by their buttonlike disks and scalloped petals. We have two species, *H. autumnale*, and *H. nudiflorum*, both of which grow in wet meadows. The name sneezeweed is a false charge. No showy flower designed to be pollinated by insects is responsible for allergies. This is the fault of wind-pollinated plants that bloom at the same time, such as ragweeds and grasses.

The genus *Bidens* includes several showy species as well as some that go unnoticed until they produce the two-pronged seeds called "beggar's ticks" or "Spanish needles." One of the showiest is *Bidens polylepis*, the tickseed sunflower, a midwestern species that has invaded the mid-Atlantic states in abundance, especially on the Coastal Plain, where it turns roadside ditches and low-lying fields into a sheet of brilliant yellow. It is an annual that self-seeds abundantly.

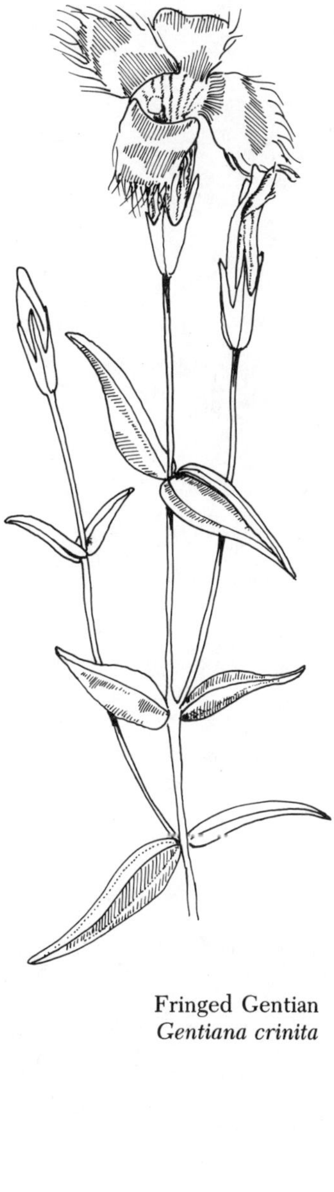

Fringed Gentian
Gentiana crinita

Last in the meadow to bloom are the gentians, named for a King Gentius who ruled Illyria (now Yugoslavia) about two thousand years ago. He is supposed to have used the roots for medicine. With one exception, all eastern gentians are a deep, clear blue. Some, such as the bottle gentian, *Gentiana andrewsii*, and the closed gentian, *G. clausa*, have self-fertile flowers that never open. The most famous species, the fringed gentian, *G. crinita*, has now become very rare. It is a true biennial that takes two years to bloom and then dies. If a field where fringed gentians grow is moved at any time during the summer, the flowering stalk is cut and the plant dies without setting seed. Studies have shown that fringed gentians also need a high level of magnesium in the soil and an exactly neutral pH of 7.0. So particular are their requirements, one wonders how they can survive, yet records show they were plentiful in eastern meadows around the turn of the century.

The many species of sedges and rushes are also reliable indicators of damp soil and low meadows. We will look at them in more detail in the next chapter, when we move closer to the water and examine wetlands.

7
WETLANDS

Scant oxygen is the critical environmental factor for wetland plants. Most oxygen is, of course, produced by plants as a byproduct of photosynthesis, but oxygen is also used by plants for respiration, just as it is by all living organisms except yeasts and certain bacteria. In its most basic sense, the term respiration means the release of energy from food by oxidation: energy that is necessary for growth and all the complex chemical reactions that sustain life.

The roots of land plants absorb oxygen from tiny air spaces found between the particles of a well-drained soil. In saturated soil, however, these spaces are filled with water, and the only oxygen available is dissolved in the water, but dissolved oxygen is less than 5 percent of that found in an equivalent volume of air. As a result, aquatic plants have had to develop special tissues that will store air as well as keep them afloat. Many also have a hard coating on their foliage not unlike that which protects desert plants from desiccation, only in this case the purpose is to keep air in and water out.

All terrestrial plants originally migrated onto dry land from a water environment. Aquatic plants then returned, presumably to escape competition. They did not simply stay behind. We know this because they hold their flowers above the surface to be pollinated and subsequently develop seeds. It was the evolution of seeds, with their unique capacity to store offspring potential safely sealed in a weatherproof case, that enabled plants to leave the water and colonize dry land in the first place.

The water in aquatic habitats may be fresh, brackish, or salt, with pH factors that range all over the scale. Many freshwater plants can tolerate some degree of brackishness, but those of the true salt marsh are a separate community. Peat bog communities are distinctive, too, because bog water is too acid to support the bacteria that decay organic matter, severely limiting available nutrients.

Ponds and Swamps

Natural ponds occur in valleys when water drains from surrounding hills to form sheets of standing water, or a river's current severs an "oxbow" from its own meandering course. Beavers and landslides often dam high mountain streams to create shallow ponds studded with the pale trunks of drowned trees. Yet most of the olive-brown country ponds that are edged with cattails and smeared with lily pads are artificial, dug to water cattle or support fish and waterfowl. Often a low spot by a woodland stream may be all that is left of such a pond, constructed over a century ago to provide power for a grist- or sawmill.

Every pond, whether natural or artificial, is a successional habitat. All ponds eventually grow old and die, successively becoming cattail marshes, shrub swamps, wooded swamps, and eventually dry woods. The process, called eutrophication, is natural, yet today it is often unnaturally accelerated by runoff from cultivated and developed land as well as by waste water containing phosphates that fertilize and stimulate the growth of water plants.

Lakes are successional, too, although a lake may take thousands of years to age and die. Existing glacier lakes are usually hundreds of feet deep, far too deep and cold for plants except around the edge, yet is is thought that some northern bogs and prairie potholes are actually the remains of extinct glacial lakes.

It is plants that age a pond. Advancing into the center in concentric rings, they trap silt and gradually fill the bottom with organic debris. As the pond becomes shallower, each group of plants creeps farther into its neighbors' habitat until the open water disappears altogether. This process begins immediately upon the birth of a pond, as thousands of seeds

Marsh Marigold
Caltha palustris

blow in on the wind or float down a feeder stream while hundreds more hitch rides on the muddy feet of wading birds.

A pond in its productive prime is called mesotrophic. We might approach such a pond through a wooded swamp, once a pond itself, but now only flooded intermittently by runoff. The water level in such a swamp may be as deep as a foot in winter and spring, yet dry up completely during a late summer drought. If the swamp is old enough, it may be well on its way to becoming a woods filled with silver or red maple, *Acer saccharinum* or *A. rubrum*, and sycamore, *Platanus occidentalis*, trees also common on flood plains where they are subject to periodic flooding. Or it may become a thicket of black willow, *Salix nigra*, alder, *Alnus rugosa*, and elderberry, *Sambucus canadensis*, perhaps sparked in winter by the brilliant red berries of *Ilex verticillata*, the winterberry, or the peculiar globular seed pods of buttonbush, *Cephalanthus occidentalis*. If the season is early spring, marsh marigolds, *Caltha palustris*, may glow like glistening yellow buttercups in mud dotted with the purple hoods of skunk cabbage, *Symplocarpus foetidus*, while in summer the air may be sweet with the fragrance of the coastal sweet pepperbush, *Clethra alnifolia*, or swamp azalea, *Rhododendron viscosum*. Swamp azalea was the very first American plant to travel back to England; seeds were sent to England by the Virginia colonists in 1680. In summer, too, wet meadow species described in the last chapter will be found on the banks of a pond.

Swamp Azalea
Rhododendron viscosum

Emergents

The shallow water close to the bank of the pond is known as the littoral zone. This is the domain of the emergents, plants with their roots anchored firmly in submerged soil, but with stems, leaves, and flowers held well above the surface. An example is the broad-leaf cattail, *Typha latifolia.* Cattails are monoecious: they have separate male and female flowers on the same plant. In this case, the male flowers are clustered at the top of the stem and fall away as soon as the female flowers that form the familiar brown cigar-shaped inflorescence are fertilized.

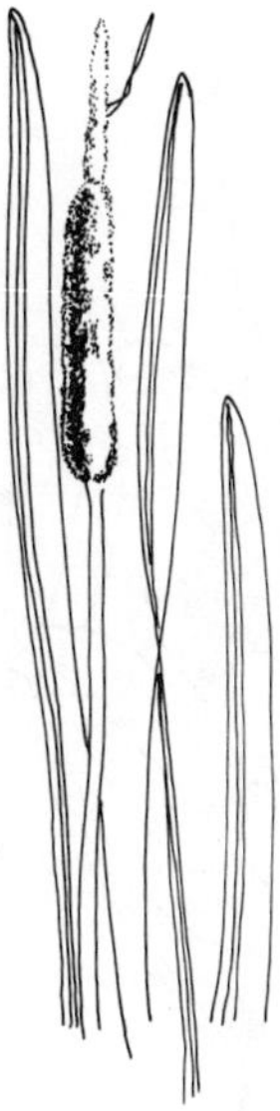

Cattail
Typha latifolia

Cattails often grow in the littoral zone with two members of the arum family, Araceae: golden club, *Orontium aquaticum,* and sweet flag, *Acorus calamus.* The individual flowers of golden club are minute, but they are arranged on a conspicuous yellow-tipped "club," botanically called a spadix. Skunk cabbage and jack-in-the-pulpit are also members of this family and have similar flowers, but their spadixes are concealed by a hood. The folk name for golden club is "never wet" because its flat leaves have a waxy coating that sheds water like a duck. Sweet flag, on the other hand, has rigid swordlike leaves, fragrant when crushed, and a spadix of tiny, tightly packed flowers that jut out at an angle from the stem. Sweet flag was called "wisuhkum" by the Leni-lenape Indians, a word meaning "it tastes bitter." The Indians mashed the root and used the juice to treat heart disease.

Blue Flag
Iris versicolor

Blue flag, *Iris versicolor*, has similar leaves yet is a very different plant. This lovely native iris blooms in shallow water in May, just before its robust alien cousin, the handsome yellow *Iris pseudacorus.* Swamp loosestrife, *Decodon verticillatus,* also known as water willow, is a summer-blooming emergent with tufts of lavender flowers in the upper leaf axils. Water willow is a shrubby plant with arching branches that root at the tips. It forms dense thickets that trap decaying vegetable matter and gradually extend the shoreline of the pond. Water willow belongs to the same family, Lythraceae, as the alien purple loosestrife, *Lythrum salicaria.*

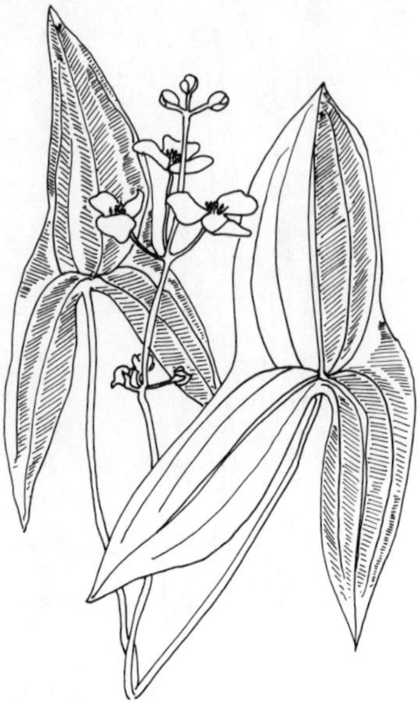

Arrowhead
Sagittaria latifolia

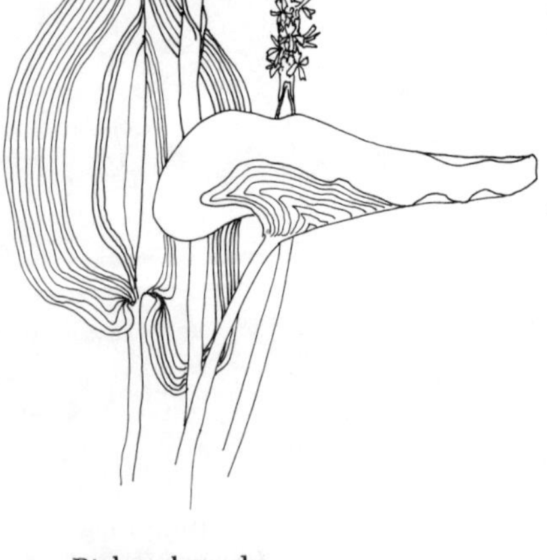

Pickerelweed
Pontederia cordata

Other plants that are often called loosestrifes, such as fringed loosestrife, *Lysimachia ciliata,* and swamp candles, *L. terrestris,* are actually members of the primrose family, Primulaceae.

Arrowhead, *Sagittaria latifolia,* is another common plant of the littoral zone. The Indians called it "duck potato" for its edible root, a golf-ball-sized tuber that ducks do, in fact, pull up and eat. The pale blue spires of pickerelweed, *Pontederia cordata,* may form large masses in shallow water at the edge of ponds, while various sedges and rushes are reflected in jagged stripes across the quiet surface.

Wool grass, *Scirpus cyperinus,* a conspicuous grasslike plant with a floppy brown inflorescence, is not a grass but a sedge. Umbrella sedge, *Cyperus strigosus,* is another species found at the edge of ponds. Most sedges have stems that are triangular in cross section ("sedges have edges") narrow leaves arranged in threes, and prefer wet places. They are not only common near ponds, but in freshwater marshes, roadside ditches, and even in poorly drained fields. The river of grass that is the Florida Everglades is actually a river of sedge, *Claudium jamaicensis.*

Some sedges are popularly called bulrushes, genus *Scirpus,* or spike rushes, genus *Eleocharis,* a practice that thoroughly confuses the issue. True rushes belong to the family Juncaceae and have stems as round as knitting needles. Two rushes commonly found on the edges of ponds are soft rush, *Juncus effusus,* and Canada rush, *J. canadensis.* The so-called chairmaker's rush is not a rush but a sedge, *Scirpus americanus.* It has a sharply triangular stem with a scaly, brown flower cluster halfway up the side.

Two grasses that may form large patches in wet, marshy spots are reed canary grass, *Phalaris arundinacea,* and blue joint, *Calamagrostis canadensis.* Reed canary grass makes excellent hay, so it is often planted on upland sites as well.

Floating Leaved Plants

Beyond the emergents of the littoral zone, but still in relatively shallow water, we come to the floating leaved plants. They have long pliable stems and round padlike leaves with no lobes or serrations that might be torn by wind or waves. The leaves are often protected by a hard, waxy coating to prevent waterlogging as well, for the water that buoys them up also thrusts their flat surfaces into the full force of pelting rain and hail.

Floating leaved plants are true aquatics and include the waxy, fragrant waterlily, *Nymphaea odorata,* as well as the coarser, more invasive pond lily or spatterdock, *Nuphar variegatum* or *N. advena.* Water shamrock, *Marsilea quadrifolia,* is a floating leaved fern that looks rather like a large four-leaf clover. It has spread throughout the East from a pond in Litchfield County, Connecticut, where it was introduced in 1862.

Umbrella Sedge
Cyperus strigosus

American lotus, *Nelumbo lutea,* is a beautiful native aquatic that is not common in the East, although it is despised as a weed on parts of the Mississippi River. Lotus flowers are a clear yellow, and the seed pod that follows them looks like a perforated brown funnel. It is often sold by florists for dried arrangements. The round leaves of the lotus are pelitate, meaning that the stalk is attached to the center, rather like an inverted umbrella.

Floating heart, *Nymphoides* spp., are delicate aquatic members of the gentian family, Gentianaceae. They have small, waterlily-like leaves and umbels of tiny, white, five-petaled flowers. *N. aquatica* is found on the Coastal Plain south of New Jersey. *N. cordata* is a smaller, more northern species.

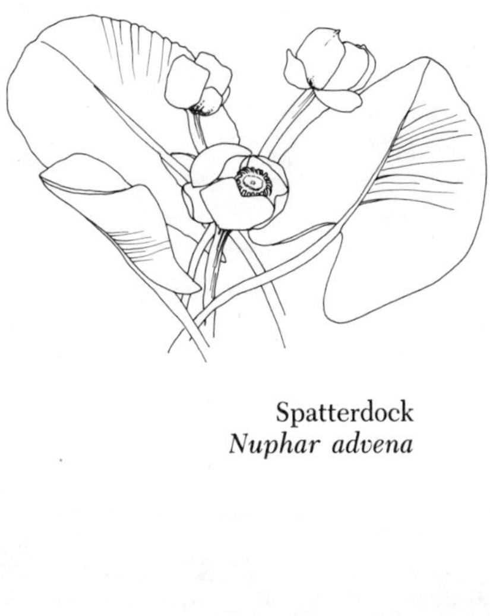

Spatterdock
Nuphar advena

Free-floating Aquatics

Free-floating aquatics have no contact with the bottom, but float in tangled masses on the surface of the water, where they are vulnerable to pounding by

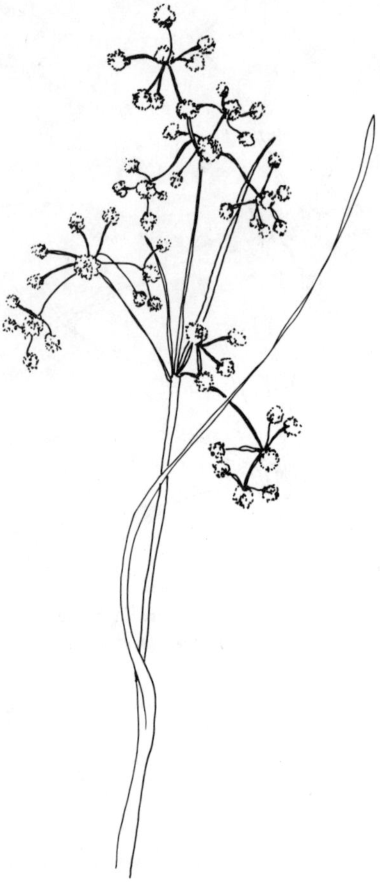

Canada Rush
Juncus canadensis

waves and swift currents. For this reason they are only found in still water, where they may form mats dense enough to clog navigable waterways. Most floating aquatics grow in a form of a rosette and multiply by means of long stems called rhizomes. Their leaves and stems are often swollen with air tissue and act like bladders to keep them afloat, while their hairy roots dangle beneath, absorbing nutrients from the water. Duckweed, *Spirodela polyrhiza,* is a common floating plant of eastern ponds. Its tiny, swollen "leaves" are not leaves at all, but modified stems. Duckweed may blanket a pond in summer, filling the space between the waterlilies like coarse green algae. In winter, it simply drops to the bottom, hiding beneath the ice until the warm water of spring causes it to float to the surface again.

Water hyacinth, *Eichhornia crassipes*, is a floating aquatic with beautiful blue flowers and glossy, deep green foliage, but it is a notorious weed of Coastal Plain waterways south of Virginia. We know that specimen plants of water hyacinth, a native of South America, were given as souvenirs at an International Cotton Exposition held in New Orleans in 1884. They were soon clogging river channels all over the south. Happily, it has been discovered that water hyacinths can purify polluted water, so they are now being put to good use in sewage and factory-waste lagoons.

Freshwater Tidal Marshes

All the plants discussed so far grow in nontidal waters of inland ponds and rivers, yet the influence of the tide can also be felt along the margins of coastal rivers and bays many miles from the sea. A conspicuous example is the Chesapeake Bay and its tributaries, where a wedge of brackish water develops as the heavier salt water sweeping in on the tide sinks beneath fresh water flowing down the rivers. This brackish wedge not only moves about on the tide, it also travels up or down stream from season to season according to the amount of rainfall. Thus, the marshes that develop in coves along estuarine rivers may grow in water that is sometimes fresh, sometimes brackish. Many pond plants, such as arrowhead, cattails, and pickerelweed, can tolerate such minor fluctuations of salt content and often grow in river marshes. Soft rush, *Juncus effusus*, is found here as well, often growing with arrow arum, *Peltandra virginica*, a plant with the distinctive spadix of the arum family and large, triangular leaves. The smartweeds, genus *Polygonum*, also thrive here. Three species often found are *P. punctatum*, *P. densiflorum*, and *P. hydropiperoides*. All have tiny pink or white flowers that are clustered in a terminal spike. Characteristic of smartweeds is a tissuelike sheath that surrounds the nodes on the stem. Halberd-leaved "tear thumb," *P. arifolium*, is a pesky member of the same genus. It has sharp, downward-pointing barbs all along its stem and drapes itself over other marsh plants like a spiny fishnet. Northern wild rice, *Zizania aquatica*, is a valuable grass of estuarine marshes, with grains that are a gourmet treat for both man and waterfowl. The plant is distinctive because the male flowers rise on a plumelike inflorescence at the top of the stem while the female flowers fan out beneath.

Arrow Arum
Peltandra virginica

In late summer, rose mallows decorate these marshes with enormous pink or white flowers that look as if they were fashioned from crepe paper. *Hibiscus moscheutos* may be pink or white, but each blossom has a center of deep carmine, while *Hibiscus palustris* is pure pink and more common in New England. Rose mallows are sturdy plants with large, heart-shaped leaves. They have sculptured buds and pointed bracts that surround the developing seed pods like a cage. Sometimes the frosty pink heads of *Mikania scandens* twine among the mallows. It is the only climbing member of the Composite family and has foliage rather like a morning glory.

Rose Mallow
Hibiscus palustris

Climbing Hemp
Mikania scandens

As fall approaches and the rose mallows fade, two species of tickseed sunflower paint some river marshes a brilliant yellow. They are *Bidens coronata*, which has rather delicate, deeply lobed leaves, and *B. laevis*, with thick, almost succulent, opposite leaves. Towering twelve or more feet above them may be a single stately plant of narrow-leaved sunflower, *Helianthus angustifolius*.

Many river marshes, particularly where the natural water level has been artificially altered, have been invaded by *Phragmites communis*, the giant reed. A member of the grass family, Gramineae, phragmites can form impenetrable stands, to the exclusion of all other plants. It has become an unwelcome symbol of the urban Northeast, appearing in roadside ditches, dumps, industrial parks, or wherever the drainage is poor. A conspicuous example is the so-called "Jersey Meadows," a marshy plain covering thousands of acres just across the Hudson River from New York City. It is sad to think that these meadows were once covered with a variety of freshwater marsh plants and, according to an early New York botanist named John Torrey, included a white cedar swamp full of rare orchids. "No region is more interesting to a botanist . . . than that which surrounds the City of New York," he wrote in 1819. The reeds apparently began to invade in 1867 when dikes were built that killed the swamp.

Viewed objectively, phragmites is a handsome plant, with large terminal panicles of silky purple flowers that turn gray in the fall and winter. Unlike most weeds of modern civilization, it is not an alien, but a plant of worldwide distribution. Remains preserved in ancient peat bogs show it has been on this continent for thousands of years, and there is evidence that the Indians used it for arrow shafts well before the coming of the Europeans.

Phragmites needs a wet soil with a neutral pH, so its rapid spread often indicates that sewage pollution, which is alkaline, has altered the natural acidity of the water. Many stands are holdovers from the days when dredge spoil from ditches and canals was commonly dumped onto salt marshes, raising the original level of the land. The reeds cannot tolerate fluctuating water levels such as those that occur with the natural rise and fall of the tide, so the key to eliminating them seems to lie squarely on intelligent land management. Mowing or burning merely encourages stronger growth, and herbicides work only temporarily. Most herbicides cannot be used near water courses in any case, for they have serious side effects.

Salt Marshes

A true salt marsh is a fluid plain of meandering creeks, black mud, and grass: a transitional zone between land and sea. These marshes typically develop behind barrier islands where silt pushed down rivers mixes with sand washed inland by the sea. These sediments are quickly invaded by salt-tolerant plants, many of which have cordlike roots that sift and bind them, gradually extending the land. Extensive salt marshes are found on the Coastal Plain, where a small rise in sea level will flood the low-lying land with salt water as far as the tide can reach. New England marshes are found on the coastal outwash plain, those heaps of sand and gravel that were left behind by melting glaciers.

Few species of plants can grow in the hostile environment of a salt marsh. Paradoxically, although it is wet much of the time, a salt marsh is a physiologically dry habitat for plants, because salt concentrations in the surrounding water cause living cells to lose water by osmosis. Without special adaptations, plants in such an environment would soon shrivel and die.

For instance, many species of *Spartina*, the principal genus of grass on the salt marsh, have developed a membrane on their roots that excludes salt, while at the same time cells in other parts of the

Salt Hay
Spartina patens

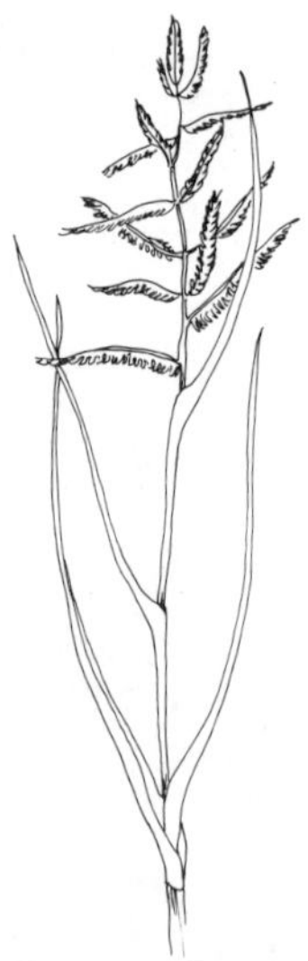

Spartina cynosuroides

plant actually allow the salt to concentrate in order to counteract osmotic pressure. *Spartina* even has glands on its leaves that excrete salt. Yet in spite of all this, experiments have shown that not only does *Spartina* grow luxuriously in fresh water, its seeds will only germinate in fresh water, indicating that the plants must have moved onto the salt marsh in order to escape competition from more robust freshwater plants.

There are three species of *Spartina* found in eastern salt marshes, each dealt a particular habitat by the tide. At the seaward edge and along the fingerlike creeks that invade the marsh, where the mud is black, sticky, and constantly flooded, we find *S. alterniflora*, the saltwater cordgrass, *S. alterniflora* is the colonizer of the marsh, with strong underground rhizomes that trap silt and extend the land. On the southern Coastal Plain, where large areas of marsh are flooded daily by the tide, it is the dominant grass. On the higher salt marshes north of Virginia, *S. patens*, the salt hay, colonizes that part of the marsh that is only reached by spring tides that occur with the full and new moon. The third species of *Spartina*, *S. cynosuroides*, grows along the marsh's landward boundary where the water is less salty, safely out of reach by all but the highest storm tides. *Spartina cynosuroides* is a tall, handsome grass that turns a pale ocher in fall and winter. It is not found north of Connecticut.

The *Spartina patens* of the high marsh has a tidy, mowed look and is usually dry enough to walk upon, in fact, cattle were grazed here in earlier days and the grass was frequently mowed for fodder. Spike grass, *Distichlis spicata*, grows in more waterlogged areas, and interwoven with both are drifts of the delicate black rush, *Juncus gerardi*, replaced farther south by the species *Juncus roemerianus*. These are all short, slender plants and their varying colors blend and mix in a subtle abstract design, especially beautiful in autumn when they turn yellow ocher, burnt sienna, and umber.

In among the grasses are bare areas called salt pannes, where evaporation has apparently caused the salt to accumulate in high concentrations. Here, only *Salicornia* will grow. It is a true halophyte, or salt-seeking plant, that will not grow in a freshwater environment. *Salicornia* has cylindrical fleshy stems, green in summer, carmine in autumn. It is a member of the goosefoot family, Chenopodiaceae, and, like most other members of that family, is edible. Country people used them to give a salty tang to salads and pickles and called them "chicken toes," "glasswort," or "saltwort."

Bayberry
Myrica pensylvanica

No woody plants grow on the open marshes of the East Coast as mangroves do in Florida, and some botanists think that is because the ice that glazes the marsh in winter would sever woody seedlings. However, three woody shrubs do grow along the landward edge, colonizing the intermediate zone between marsh and dry land where they are only touched by the highest storm tides. They also cluster on high spots in the open marsh forming dark islands in a sea of grass. Both bayberry, *Myrica pensylvanica*, and its more southern relative, sweet gale, *M. cerifera*, grow here. So does *Iva frutescens*, the marsh elder, an awkward shrub with the typical fleshy leaves of plants growing in a maritime environment. *Baccharis halimifolia*, the high tide bush, is a woody member of the composite family, inconspicuous until early fall when the developing seed heads look like hundreds of tiny white plumes. American holly, *Ilex opaca*, fills the woods that grow along the borders of salt marshes on the Coastal Plain, often mixed with two relatives, inkberry, *I. glabra*, a shining evergreen shrub that is much used in the nursery trade, and the brilliant red winterberry, *I. verticillata*.

Marsh Elder
Iva frutescens

Flowers of the salt marsh meadow are few. Most do not appear until late summer or fall. One of the earliest to bloom is salt marsh milkweed, *Asclepias lanceolata*. It is not common but may sometimes be found growing along the narrow rim where marsh

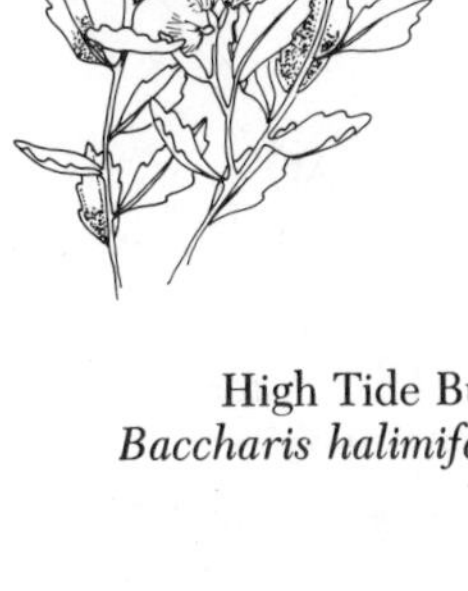

High Tide Bush
Baccharis halimifolia

Salt Marsh Fleabane
Pluchea purpurascens

Salt Marsh Aster
Aster tenuifolius

and upland meet south of New Jersey. Salt marsh milkweed has flowers of a brilliant orange, reminiscent of butterfly weed, but although the plant is slender and more delicate, the individual flowers are large. Another choice but uncommon plant of the salt marsh is the fenrose, *Kosteletzkya virginica*, a delicate pink member of the mallow family, Malvaceae. With its characteristic bushy column of fused satems and pistil, it looks like a smaller version of the robust rose mallow of brackish marshes. It is the sole northern representative of a large tropical genus that is named after a Czech botanist, B. F. Kosteletzky, and is not found north of Long Island.

By midsummer, the tiny pink seaside gerardia, *Agalinus maritima,* hides its pink bells among the salt hay, often unnoticed until nearly trod upon. Sea lavender, *Limonium nashii,* has sprays of tiny lavender flowers, each only about an eighth of an inch in diameter, but so numerous as to give the entire marsh a purple haze in late summer. Salt marsh fleabane, *Pluchea purpurascens,* is bolder, forming fuzzy pink masses along the creeks that lace the marsh. In September, salt marsh aster, *Aster tenuifolius,* appears. It is a slender, succulent plant with smooth, narrow leaves and delicate white or lavender blossoms.

Late fall brings the goldenrods. Two that grow in salt marshes are the succulent and showy seaside goldenrod, *Solidago sempervirens,* and Elliots's goldenrod, *S. elliotsi,* a slender plant with smooth, elliptical leaves that hug the stem.

Bogs

Streams bubbling into a pond, tides gliding over a salt marsh, river currents sweeping seaward—all renew the dissolved oxygen supply and bring fresh silt rich in nutrients for use by wetland plants. A bog, however, has no flowing water, very little oxygen, and no silt. Bogs—sterile, soggy, acidic, and characterized by quaking mats of vegetation—are the typical wetland community of northern latitudes. Here, where the cool climate inhibits evaporation, bogs spread like watery mantles over shallow depressions originally left by the glaciers. Depressions are all that remain of ancient lakes that have slowly filled with peat over thousands of years.

The podzol soils of northern forests are typically sandy and too acid to readily form humus. In fact, sandy soils, wherever they occur, tend to be deficient in soil bacteria and thus conducive to bog formation in areas of poor drainage, particularly near the coast where fog cools the summer climate. In southern New England, Cap Cod, and offshore islands, such as Martha's Vineyard and Nantucket, where the gravelly soil was originally deposited by glaciers, such bogs develop in depressions called "kettleholes."

South of the boundary that once separated glaciated land from unglaciated, bogs are less common, yet they do occur in poorly drained depressions in the Appalachians where complicated geologic formations have resulted in impermeable soils and perched water tables. They also occur in the sandy soil of the New Jersey Pine Barrens where the aquifers are close to the surface. All these bogs contain northern plants that were pushed southward by the glaciers.

Savannas are shallow ponds on the sandy Coastal Plain where the soil is acid and the drainage poor. They develop bog communities dominated by grasses and sedges. Because numerous species of pitcher plants, genus *Sarracenia*, are found in such bogs along the Gulf of Mexico, they are thought to be the ancestral home of S. *purpurea*, the only species found in the north.

The bald cypress, *Taxodium distichum*, hovers over wooded swamps and bogs in the South. Their spreading limbs, draped with the ghostly fingers of Spanish moss, seem to loom above the black, tannin-filled water. Roots as big as fire hoses penetrate deep into the mud and send up buttresses, called "knees," to surround the swollen base of the main trunk like a palace guard. The purpose of these "knees" is not precisely known, for the old idea that they supplied oxygen to submerged roots has not been supported by anatomical research. Bald cypress is a gymnosperm of the family Taxodiaceae, a stately group that also includes the California redwoods. Unlike most conifers, it is deciduous. An ancient specimen may be over a thousand years old and a hundred and fifty feet tall, although such venerable trees are rare because most of their habitats were logged out during the early years of this century. Spanish moss, *Tillandsia usneoides,* is not a true moss but a flowering plant of the bromeliad or pineapple family, Bromeliaceae. It is epiphytic, which means it grows upon another plant for support, but it is not a parasite. Spanish moss propagates by simply being broken. Pieces are carried on the wind until they hit a snag, whereupon they attach themselves and begin to grow.

Because the aerating effect of flowing water is absent, bog plants must subsist on oxygen levels even lower than that of other wetland plants. In addition, many bogs have a pH factor as low as 3.5, only half a point higher than vinegar. The pH factor is calculated logarithmically, so each point on the scale is ten times higher or lower than the one preceding it. Therefore, a soil with a pH of 3 is ten thousand times as acid as one registering a neutral pH of 7, the preferred range of most plants. The varied species of soil bacteria responsible for the decomposition of organic material cannot live in such an acid environment, so dead bog vegetation does not decay but rather accumulates as peat. With no bacteria, peat bogs are so sterile that lumber from fallen trees is usable after centuries of immersion, and ancient seeds removed from bogs sprout when planted. Even bodies have been retrieved undecomposed, and sphagnum moss has effectively prevented infection when used to dress the wounds of soldiers.

This talent for preservation makes bogs particularly interesting to plant historians. Each minute pollen grain that is released over the land every summer has a tough outer coat. When encased in layers of sterile peat, this shell has proven so durable that it is possible to identify the exact species that produced the pollen thousands of years ago. Thus core studies of accumulated peat at the bottom of bogs reveal when vegetation patterns have changed, indicating major shifts of climate.

Buoyant, quaking mats of absorbent sphagnum moss are the characteristic feature of northern bogs. That these mats are also prevalent in the relict bogs of southern New England, the Pine Barrens, and the Appalachians tells us that these are northern communities that were originally forced south by the glaciers. Sphagnum is less common in the grass-sedge bogs of the south, where soil acidity is caused by mineral compounds, not accumulating peat.

Sphagnum is a true moss, or bryophyte, and thus lacks roots, true stems, or leaves. It forms a tangled, pale green cushion, sometimes up to a foot thick over the surface of the bog. The distinctive color results from special cells that lack chlorophyll but are able to absorb up to sixteen times the plant's own weight in water. It is this ability to retain water, plus the moss's inherent sterility, that makes sphagnum such a valuable tool for horticulture, as seeds started in it stay moist but do not "damp off" from fungus disease. The peat moss widely used as a soil conditioner is actually dead sphagnum.

Most plants of the open bog grow directly in this sphagnum cushion, adapting to the acidity and lack of soil nutrients in a variety of curious ways. The most obvious examples are the carnivores, a peculiar group of "insect-eating" plants of which the pitcher plant, *Sarracenia purpurea*, is often the most conspicuous.

Pitcher plants have thick, rubbery, tube-shaped leaves with a ruffled lip. These tubes, or "pitchers," collect rainwater. Glands on the lip secrete a nectar

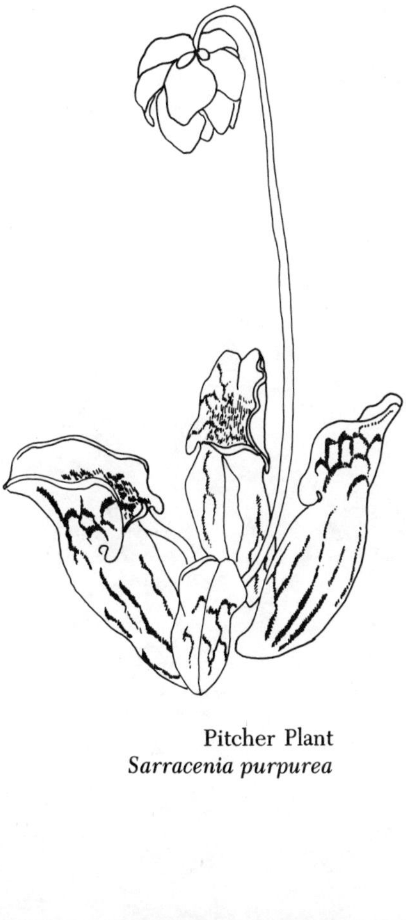

Pitcher Plant
Sarracenia purpurea

that attracts flies like carrion does, but a fly that ventures inside is immediately doomed. Short, bristlelike hairs around the lip ease his way downward, but effectively prevent him from ever climbing out again. The fly drowns in the rainwater at the bottom of the tube, and the nutrients released by the decomposition of his body are absorbed by the plant's tissues.

In June, the pitcher plants bloom, each flower appearing on a separate leafless stalk. The deep red-purple petals are short-lived, dropping after a day or two, but the curious umbrella-shaped style, or female organ, surrounded by a leathery calyx, remains on the plant all winter, giving it the look of an ingenious red-brown pinwheel.

Sundew, *Drosera* spp., is another insectivorous plant that is less conspicuous but more widespread than pitcher plants, occurring on acid, sandy soil throughout the Coastal Plain as well as in northern bogs. Most sundews form tiny rosettes of spoon-shaped leaves, although one Coastal Plain species, *D. filiformis*, has leaves like slender hairy caterpillars. This plant was one of several North American species included in Charles Darwin's book *Insectivorous Plants*, published in 1875. The leaves of all sundews are tipped with sticky glands that attract gnats and midges, snaring them like flypaper. The glutinous hairs then wrap themselves around the captive and digest him. Sundews are annuals and have tiny white, pink, or purple five-petaled flowers in summer.

A third carnivore, bladderwort, *Utricularia* spp., forms a mat of fine, threadlike leaves just below the surface of the water. Attached to them are minute round bladders, the size of a pinhead, with lids that close on pond insects like tiny trap doors. The flower, which superficially resembles a snapdragon, blooms in June, rising above the water on a wiry stem less than a foot tall. The color may be yellow or pink. There are many species of bladderwort in the East; some grow in the still water of lakes and ponds, while others prefer bogs.

Perhaps the most famous carnivore of all is the Venus' flytrap, *Dionaea muscipula*, a plant that snaps at its prey like an animal. The instant an insect touches the hairlike triggers on the upper surface of the leaf, the two lobes snap together, snaring the hapless victim, whose struggles thereafter only serve to stimulate the production of digestive juices. The leaf stays closed for several days until all the soft parts of the animal are consumed. Venus' flytrap only grows in a small area within fifty miles of Wilmington, North Carolina, and its populations have been seriously threatened by unscrupulous and unlawful collecting.

Contrary to popular ideas, carnivorous plants do not live on insects in the same way carnivorous animals live on their prey. All these plants contain chlorophyll and carry on photosynthesis as their primary source of energy. The bodies of the insects merely supply nutrients missing in the acid environment of the bog and thus act in much the same way as a fertilizer.

Throughout the world the orchid family, Orchidaceae, is one of the largest families of flowering plants. Most orchids are tropical; of those that do grow in temperate climates most are found in bogs. The flowers of orchids range from the inconspicuous to the flamboyant, but all have a bilateral arrangement of three petals with one, called the "lip," differing from the others in size, shape, or color and sometimes in all three. Yet this is not always easy to sort out, for the sepals of orchids often resemble petals and may combine with the lateral ones to form a hood. A center column, fused from the style and stamens, is also typical of orchids, but this too may be brightly colored and resemble a petal. A further peculiarity of orchid flowers is their propensity to

Pink Lady Slipper
or Moccasin Flower
Cypripedium acaule

Rose Pogonia
Pogonia ophioglossoides

twist on their stems during development, so that the "lip" may start out as the top petal but end as the lowest.

Platanthera, a genus containing the rein and fringed orchids, contains most of the species found in the bogs of northern New England, but lady-slippers, *Cypripedium* spp., (which are also found in moist, acid woods), rose pogonia, *Pogonia ophioglossoides*, swamp pink, *Arethusa bulbosa*, grass pink, *Calopogon tuberosus* and ladies' tresses, *Spiranthes* spp. may occur, too, as well as in more southern bogs.

Orchids have adapted to the bog environment in quite a different way from the carnivores. A group of fungi species called *mychorrizae* live symbiotically on their roots, as they do on many acid-loving plants. These fungi act as stand-ins for soil bacteria, decomposing organic matter to make essential nutrients available to the plant. This delicate relationship is the reason why so many orchids are impossible to transplant, although thoughtless gardeners cannot seem to resist digging them up whenever they find them. Alas, when disturbed, the fungus invariably dies, and the orchid soon follows.

Cotton grass, *Eriophorum virginicum*, is a sedge with a tuft of silky hairs on the tip of a wiry stalk that dots the bog's sphagnum mat in summer. It is a plant of the Arctic Tundra that reaches the southern limit of its range in the bogs of the New Jersey Pine Barrens. Bog bean, *Menyanthes trifoliata*, is a creeping plant of cool northern bogs that spreads throughout the wet peat. It has a six-inch stem tipped with a raceme of delicate white flowers, each with a brush of glistening hairs on the upper surface of the petals. The roots of bog bean are used in Scandinavian countries to make a starchy cake called "missen bread."

Pipewort, or hatpins, *Eriocaulon septangulare*, is another northern plant that forms sedgelike tussocks in the bogs of the Coastal Plain. The minute white flowers cluster in a tight sphere at the tip of a thin stem which indeed gives the plant the look of an old-fashioned hatpin. *E. decangulare* is a similar southern species. *Rhexia* spp., the meadow beauties, are southern plants that occur in sandy bogs throughout the Coastal Plain. They have delicate pink, four-petaled flowers with conspicuous yellow stamens. The seed pods are shaped like tiny urns.

Orange milkwort, *Polygala lutea,* looks like bright clumps of orange lollipops in the sphagnum mat of Coastal Plain bogs. The blossom is a tight head of tiny flowers, rather like a clover, and extremely long-lasting, as it continues to bloom from the center as the faded petals drop from the outer rim. Milkwort is a biennial, so for a colony to persist, conditions must be right for it to reseed itself.

Extremely rare, but intriguing, are the bog bunchflowers, primitive members of the lily family, Liliaceae. Tradition says that the lilies growing in the Elysian fields, that paradise of happy souls of Greek mythology, were bunchlilies. If so, these are some New World relatives. Two species, representing two genera, grow in southern bogs: the bog hyacinth, *Helonias bullata,* and the bog asphodel, *Narthecium americanum.* Bog hyacinth has tightly clustered flowers of a dull pink, each with blue stamens, that rise on a stout stem from a basal evergreen tuft of wide, straplike leaves. Staten Island apparently represents the northernmost limit of this plant on the Coastal Plain. It also grows in mountain bogs from Pennsylvania into Georgia. The bog asphodel has yellow flowers, also bunched at the top of the stalk, and thin, grasslike leaves. It grows in an isolated community in the New Jersey Pine Barrens and is not found again until the mountain bogs of North Carolina. Such a widely disjunct community usually means a plant of ancient lineage, a survivor of another age.

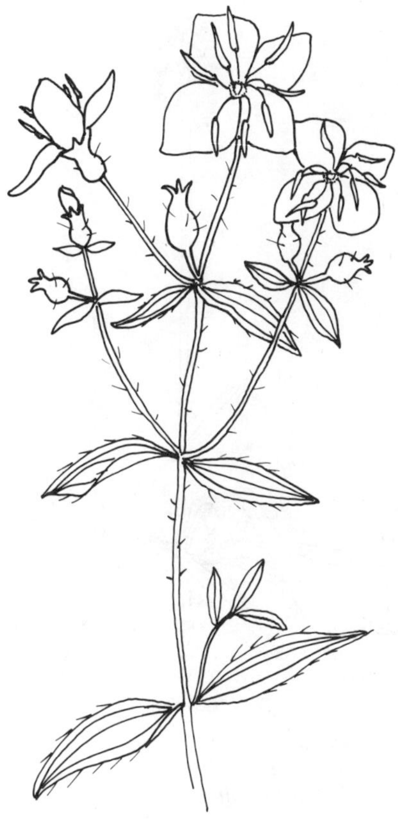

Maryland Meadow Beauty
Rhexia mariana

The shrubs that surround a bog, sometimes forming islands in the sphagnum mat as well, are

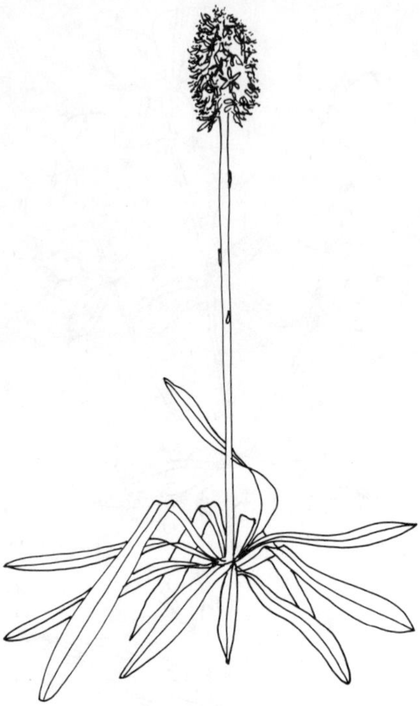

Swamp Pink
or Bog Hyacinth
Helonias bullata

predominantly members of the heath family, Ericaceae. Heaths are boreal plants, at home in the acid podzol soil of the northern coniferous forest, where they grow on upland sites as well as in bogs. South of the boreal belt, however, such cool acid conditions exist only in bogs, where these shrubs ring the edge in remnant communities of a once colder time.

Leatherleaf, *Chamaedaphne calyculata,* is a low, spreading shrub common around bogs. It is more or less evergreen with narrow, elliptic, minutely scaley leaves with a yellowish brown cast to their undersides. The flowers are small and white with the typical bell shape of many heaths. With it grows the Labrador tea, *Ledum groenlandicum*; its specific name, meaning "of Greenland," hints of an arctic distribution. It too is low-growing, seldom reaching three feet, and has terminal clusters of white flowers in May and June. The leaves, which are smooth on top and brown and fuzzy below, are borne on twigs that are covered with tawny hairs.

Bog rosemary, *Andromeda glaucophylla,* is a choice plant of northern bogs. It forms a compact gray-green mound covered with frail pink bells in May. Bog laurel, *Kalmia polifolia* is a rather straggly shrub with sparse purple-rose blossoms of a shape immediately recognizable to everyone familiar with mountain laurel. It leaves are white on the undersides. Lacing the sphagnum mat beneath these shrubs are the cranberries, *Vaccinium macrocarpon* and *V. oxycoccos.* Cranberries are in the same genus as blueberries yet appear quite different. They are tiny creeping evergreens that grow right in the moss. The flowers have four recurved petals and the stamens hug the style like the bill of a crane. In fact, the name cranberry is thought to be a corruption of craneberry. When seen in the wild, the familiar acidulous red berry looks much too large for the rest of the plant. Cranberries are extensively cultivated on Cape Cod and in the Pine Barrens and are among the few commercial crops native to North America. Others are blueberries, black walnuts, and pecans.

Shrubs of the heath family typically have leathery leaves, often with rolled edges. They may be fuzzy on the underside, like Labrador tea, or have a waxy bloom, like bog rosemary or bog laurel. Both attributes guard against water loss, because in spite of the soggy environment, the extremely acid water of the bog is not readily absorbed by plants.

Tamarack, *Larix laricina*, and black spruce, *Picea mariana*, are trees with worldwide distributions, occurring throughout the northern coniferous forest, but like the ericaceous shrubs, they tend to congregate around bogs in the southern part of their range. Tamarack is a deciduous conifer, dropping its needles in winter and sprouting new pale green ones every spring. It is not a valuable timber tree, whereas black spruce is logged extensively for pulpwood over much of its range. A viscous black resin exuded from the bark of this tree was once known as the lumberman's chewing gum.

These tree species are replaced in the bogs of southern New England and the Coastal Plain by the Atlantic white cedar, *Chamaecyparis thyoides*. While cedars are slow growing and may live nearly a thousand years, but in most places they have been extensively cut. Because the wood is both light weight and decay-resistant, it has been used since the eighteenth century for shingles. Peter Kalm wrote in 1749 that the lumbermen of southern New Jersey were exploiting the white cedar to such an extent that he feared it would soon be completely gone from the country.

All bogs are successional communities, gradually becoming wooded swamps and, eventually, woods. In the cool northern climate this process takes hundreds of years, but in the South it wouldhappen very quickly were it not for fire. Summer lightning fires were once common in southern savannas, sweeping across the dry surface grass in undulating line of flame that checked the growth of woody plants but left the roots of bog plants safe in the moist peat. Where these fires have been controlled, the bogs are being invaded by thickets of wax myrtle, *Myrica cerifera*, inkberry, *Ilex glabra*, and sweet bay, *Magnolia virginiana*. The increased transpiration rate of these woody plants slowly reduces the moisture in the bog, causing the bog plants to disappear.

8
DRY LANDS

A beach is built by the sea. When a wave washes soil from the shore, the roiling motion of the water floats out the lighter silt and carries it seaward, but the heavy sand is dropped as the wave breaks. A sand bar begins to form, emerging at low tide as a sand spit. The waves continue to break over it, piling up more and more sand. When the sand dries in the sun, the wind takes over, swirling the grains into patterns and sweeping them along in a moving film just above the ground. As soon as this film of sand encounters a piece of driftwood or a rock, the dune is born.

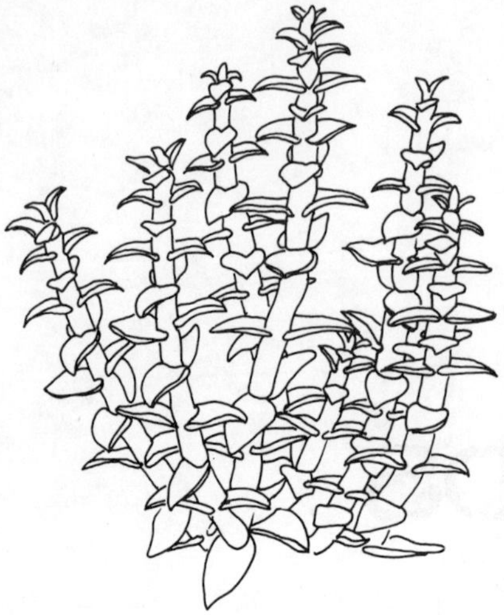

Sandwort
Arenaria peploides

Dunes

A sand dune is at the mercy of the wind that builds it, wears it down, and builds it again. Whatever steadiness it ever acquires is imposed on it by the dune grass, *Ammophila breviligulata*, which binds it like Gulliver with long underground stems called rhizomes. These rhizomes put down anchoring roots and send up clumps of waving grass every few feet, gradually carpeting the landward side of the dune and venturing over the crest to a point just beyond the reach of storm waves. Being buried by blowing sand seems only to stimulate the grass to grow; its leaves sometimes reach through as much as a yard of sand to the light. The waving clumps of grass moderate the force of the wind and insulate the surface of the dune form the blazing sun, a fact well known to children who scamper barefoot across a burning beach to cool, grass-covered dunes.

In spite of ample rainfall and moisture-laden fog, growing conditions on a sand dune approximate those of a desert. The porous sand holds little water and the summer combination of hot sun and wind constantly threatens plant tissues with desiccation. Consequently many beach plants look like desert plants. They have succulent stems and leaves with surfaces as hard and shiny as plastic: evolutionary stategies that conserve water and slow evaporation.

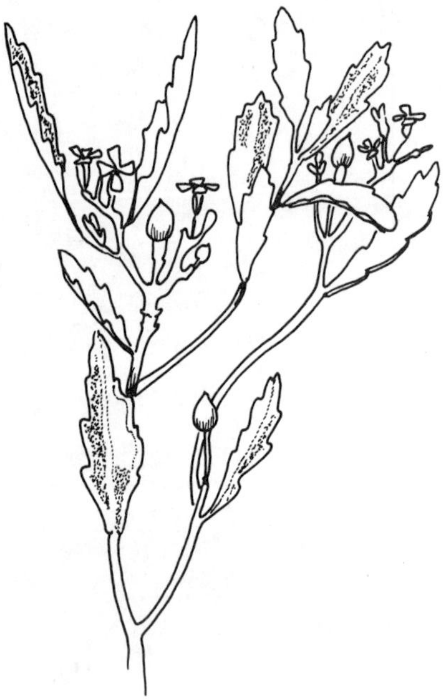

Sea Rocket
Cakile edentula

For instance, the foliage of seaside goldenrod, *Solidago sempervirens,* is noticeably smoother and thicker than its landlocked relations. Sandwort, *Arenaria peploides,* a beach-loving member of the pink family, Caryophyllaceae, forms fat clumps of rubbery leaves and stems on Atlantic beaches, producing tiny flowers in the leaf axils in early summer. Sea rocket, *Cakile edentula,* is a succulent member of the mustard family, Cruciferae. It is an annual with inconspicuous purply pink flowers in late summer and seed pods that are shaped like tiny rockets. When the seeds are ripe, they roll into the water, sometimes floating for months before washing up on another beach.

Beach Clotbur
or Sea Burdock
Xanthium echinatum

Dune plants are often prickly, too. Beach clotbur, *Xanthium echinatum,* has seed pods as spiny as a hedgehog. So does the jimson weed, *Datura stramonium,* an alien weed of dry pastures that also frequents beaches. Prickliest of all is the salt spray rose, *Rosa rugosa,* a plant that forms large colonies on the landward side of dunes in New England, where its thorny stems, wrinkled matte foliage, and blowsy flowers are a familiar sight in summer. The flowers, which come in all shades of pink as well as white, are followed by bright orange hips like cherry tomatoes. *Rosa rugosa* is a native of Asia. It arrived, so the story goes, after a ship carrying seeds and plants for an orchardist in Boston was wrecked off Cape Cod in the nineteenth century.

Halophytes, or salt seekers, of the beach include two members of the goosefoot family, Chenopodiaceae: sea blite, *Suaeda maritima,* and orache, *Atriplex patula.* The fact that these two rather unprepossessing plants have common names at all probably means that they were once gathered for food. Most members of the goosefoot family are edible: more familiar examples are beets, spinach, and swiss chard. *Salsola kali,* or saltwort, is an annual halophyte whose seeds apparently float in on the tide, for the plant appears in late summer as a dense

Salt Spray Rose
Rosa rugosa

stand just above the wrack. It is a stiff plant that is defensively prickly: each pale, waxy-green, succulent leaf ends in a sharp thorn.

Beach Pea
Lathyrus maritimus

Other plants have other stategies. Both beach pea, *Lathyrus maritimus*, and dune grass fold their foliage on dry, windy days. Beach pea clasps its leaves along the stem like so many praying hands, deflecting the rays of the burning sun, while those of dune grass fold along the midrib, closing pleatlike ridges on the inner surface against the loss of moisture. Such plants as beach wormwood, *Artemisia stelleriana*, called dusty miller in the North, and *Centaurea cineraria*, called dusty miller in the South, are covered with minute white hairs that both reflect the hot sunlight and provide insulation for the leaves. Both these plants are naturalized aliens, the first from northeast China, and the second from the Mediterranean.

The gray-green foliage of horn poppy, *Glaucium flavum*, superficially resembles that of dusty miller but sends up flowering stalks that are crowned by large yellow poppies in summer. The plant's common name comes from the long sickle-shaped seed pods that follow. Horn poppy is also an alien and seems to prefer stony beaches to sandy ones.

Beach Wormwood
or Dusty Miller
Artemisia stelleriana

Beach heather, *Hudsonia tomentosa*, forms gray-blue mats like juniper with needlelike leaves that cling to the stem like scales. No juniper, however, has beach heather's tiny yellow starlike flowers that bloom from May to July and proclaim it a member of the rock rose family, *Cistaceae*.

In their quest for moisture, most dune plants have wide-ranging root systems. They may be tap roots, spreading rhizomes, or deep-penetrating tangles like fine wire. The tops are often prostrate. Plants, such as seabeach knotweed, *Polygonum glaucum*, and the dainty seaside spurge, *Euphorbia polygonifolia*, hug the surface of the sand as if cringing from the wind. The dunes of Cape Cod, Martha's Vineyard, and Nantucket are carpeted with a low-growing plant with white woolly stems and leaves like curved stilettos. This is the sickle-leaved golden aster, *Chrysopsis falcata*. In August, it is covered with small yellow daisies that dot the sand like drops of summer sunlight.

Horn Poppy
Glaucium flavum

Behind the dunes, where the soil has acquired some humus, nestle hardy seaside shrubs such as bayberry, *Myrica pensylvanica*, and beach plum, *Prunus maritima*. Bayberry is easily recognized by the clusters of gunmetal berries along the stem in late summer and fall. The pungent aroma of its leaves is immediately reminiscent of old houses and Christmas, as bayberry candles were once made from the "bloom," or wax coating, on their surface. The candlemaker boiled the berries in a large cauldron and floated the wax off the top, but it took as much as a bushel of berries to make a single candle. Bayberry is dioecious, a term that means the male and female flowers are borne on separate plants. Obviously, only the female plants have berries.

Beach plum has two seasons of glory. In early spring, before the leaves unfold, it is a cloud of delicate white blossoms. By fall, small edible plums hang like purple pendants all along the stem. Sometimes beach plum grows right out of the dune in what appears to be a foot-high thicket, but what is in reality the top of a single bush buried in the sand. Further inland, beach plum develops into a small tree.

Bearberry, *Arctostaphylos uva-ursi*, is a plant of the far north that occurs along the Atlantic coast into New Jersey. It is common along Cape Cod roadsides and in hollows behind the dunes, where it forms large patches beneath pines pruned into bonsai silhouettes by winter gales. It has the pendulous bell flowers typical of the heaths, Ericaceae, which in this case are followed by bright red fruit. The Indians called bearberry "kinnikinnik" and reportedly smoked it like tobacco.

On undisturbed coastlines a tundralike heath may develop on high ground behind the dunes. There various species of blueberry and huckleberry, *Vaccinium* and *Gaylussacia*, mingle with pasture rose, *Rosa carolina*, shining sumac, *Rhus copallina*, and arrowwood, *Viburnum dentata* or *V. recognitum.* In August, these shrubs are interspersed with clumps of wild indigo, *Baptisia tinctoria*, a bushy plant with yellow pealike flowers. Later come the goldenrods and asters. One attractive species is the low-growing stiff aster, *A. linariifolius.* Sometimes one can find the delicate lavender-pink spires of racemed milkwort, *Polygala polygama*, in such places as well.

Barrens

The word "barrens" was bestowed by the early settlers on any area that was useless for farming because of thin, nutrient-poor soil. Botanically speaking, however, these sites are far from barren, for they support some of the most interesting plant communities in the East.

Pine Barrens

We have already described the deciduous forest as the climax vegetation pattern for most of the eastern United States. Yet there is a second pattern that some botanists call an alternate climax: the pine, pine and heath, or mixed pine and oak forests that occur on poor sandy soils. Such so-called "pine barrens" are usually associated with the Coastal Plain, but they may also occur in the Piedmont on steep, south-facing slopes and along sandy river bluffs, and even on thin, gravelly soil in the Appalachian mountains, the evolutionary parent of all eastern plant communities. In spite of adequate rainfall and a generally humid climate, all these locations are subject to punishing drought because of the inability of sand to retain moisture. Excessive dryness coupled with the incendiary quality of the resinous pines make fire a constant presence. In fact, this forest burns on a regular basis. If it does not, deciduous trees eventually take over, so perhaps it is not truly a climax community, but a successional stage that has been halted by fire.

Pine barrens exist all along the Atlantic and Gulf Coasts of the eastern United States, but the various species of pine differ according to the latitude. The northern pitch pine, *Pinus rigida,* is the dominant species in pine barrens of Cape Cod, Long Island, and New Jersey. New Jersey also represents the northernmost distribution of loblolly pine, *P. taeda,* and shortleaf pine, *P. echinata,* both trees that are common in barrens to the south. Longleaf pine, *P. palustris,* is found south of Virginia and is the dominant pine of the Sandhills area of North Carolina. Slash pine, *P. elliottii,* is the most southern species, occurring from South Carolina to Florida and the Bahamas.

Of all these locations, the New Jersey Pine Barrens are the most botanically diverse. The combination of lowland bog plants, found along the numerous slow-moving streams, and the upland plants of dry areas, plus the combination of northern and southern species, make it unique. So unique, in fact, that up to 12 percent of the plants that grow in the Pine Barrens are considered rare or endangered. In addition, these woods have been subjected to every human pressure typical of most pine barrens in the East.

In December of 1632, the navigator of a Dutch boat standing off the coast of Cape May wrote in his diary that the land gave off a "sweet perfume . . . [that] comes from the indians setting fire . . . to the woods and thickets, in order to hunt. . . ." This is the earliest reference to fire in the New Jersey Pine Barrens, an incongruous sandy wilderness of pine and oak occupying nearly one and a half million acres of the outer Coastal Plain. Only a few miles to the west of these woods is a network of interstate highways carrying an unbroken line of traffic north to New York City and south to the Delaware Memorial Bridge. Philadelphia is less than an hour's drive to the west and to the east is Atlantic City and its satellite resorts on the Jersey shore. Yet within this forest the population is somewhere around fifteen people per square mile.

People have always exploited the barrens, yet for the most part, they do not live there. Even those Indians who burned the woods in 1632 probably came from outside villages, although their ancestors had been hunting in the pines since the Pleistocene Ice Age. The Europeans, in their turn, pronounced these woods only fit for cutting wood and running cattle. Farms were small. Yet plots were not shifted around as they were elsewhere, for it took hard work and quantities of manure to make the sandy soil produce at all. Most farmers didn't bother, but simply burned off the surrounding woods and turned out their cattle and horses to graze. No one protested, for in the early days, most of land was still owned by the West Jersey Society in England. In the 1820s and 1830s, farming in a more traditional sense did experience a small boom when it was discovered that glauconite, also known as "greensand marl," a sediment plentiful on the inner coastal plain, made an excellent fertilizer for sandy soils. But like most Pine Barrens enterprises, it didn't last.

The few permanent citizens of the pines have always been forced to pick up a living where they could, and one way or another, this has usually meant cutting and burning the woods. Blueberries grow wild in the pines. In the days before they were cultivated and the practice forbidden, the bushes were regularly set afire because those that came back afterward always bore the best crops.

Bog iron occurs when acidic water precipitates

soluble iron salts from porous soil, and the smelting and working of bog iron was a thriving industry in the pines from the years before the Revolution into the early nineteenth century. Sparks from the forges and furnaces often set the woods on fire, and sometimes they were burned deliberately. Owners of timberland rarely lived in the barrens and charred trees that were useless for lumber could always be bought cheaply for the charcoal that was needed to fuel the iron industry.

The bog iron was largely depleted by the mid-nineteenth century, but by 1900 sparks from railway engines were igniting fires in the woods. The railroad brought the pines within reach of New York, causing the real estate developers to move in. They bought thousands of acres from the defunct iron companies to sell off in small lots. Yet few buyers actually moved to the pines, and those that did found the summers hot and humid and the mosquitoes unbearable. They didn't stay, and the promoters eventually went broke and sold the land for back taxes.

Today clearings are scattered among the pines where the limestone foundations of vanished buildings have altered the pH of the soil sufficiently to grow grass and poison ivy. Catalpa trees planted long ago still surround a vanished town called Martha, along with a disjunct community of a rare green orchid called Loesels twayblade, *Liparis loeselii.*

In 1978 most of the land comprising the Pine Barrens was put in the New Jersey Pinelands Natural Reserve, subjecting all future development to an orderly plan. Yet the fires continue to burn: set by the casual cigarette, the undoused campfire, or arson. All told, there are over a thousand fires every year, and only about ten of them are caused by lightning. Most places average one fire every ten to thirty years, but in an area called the Plains, the woods have been burned so often that the pines and blackjack oaks have formed peculiar multistemmed dwarfs, none taller than a man, that grow dense tops from a massive root crown called a stool.

The oaks that grow in frequently burned areas are fire selected: Post oak, *Quercus stellata*, blackjack oak, *Q. marilandica*, and bear oak, *Q. ilicifolia*. All have leaves so full of protective oil they are as thick as polished leather. Where the woods have escaped the flames for fifty years or more, the pines have been invaded by black, white, chestnut, and scarlet oak, with a lesser proportion of red maple, gum, and hickory.

Only three species of pine are able to sprout from the trunk after a fire, and two of them, the pitch pine, *Pinus rigida*, and shortleaf pine, *P. echinata*, are widespread in the Pine Barrens. The third is a native of Mexico. Pitch pine is a northern tree, growing from coastal Maine down the spine of the Appalachians, where it is always found on the thin, gravelly soil of steep slopes, ridges, or plateaus. Shortleaf pine is southern, a native of the Atlantic Coastal Plain. They are but two examples of numerous northern and southern taxa that meet in the pines. To understand how this happened, it is necessary to go back several million years.

The oldest soils of the Pine Barrens were laid down sometime in the Cretaceous period, about one hundred and forty million years ago, when dinosaurs walked the earth and flowering plants had just begun to mingle with coniferous species already a hundred million years old. Ancient pollen deposits from other locations in the Coastal Plain show that early oaks, heaths, bayberries, and members of the magnolia family, Magnoliaceae, already existed in the Cretaceous era, in fact, all the principal species that make up the dominant vegetation of today's Pine Barrens.

Yet the outer Coastal Plain, where the barrens are today, was to disappear and reappear ten times or more. Ancient seas washed over the land and then receded, boiling up the soil and laying down new layers of sediment to be cut by flowering water and roiled by ancient winds. About five million years ago, the land stabilized, and pollen evidence shows that by a million years ago, the climate was so warm that some trees and shrubs that grow in the barrens today were mixed with southern or even tropical species. The northern pitch pine was not present at that time. It moved southward ahead of the glaciers, which were still to come.

The sheet of ice known as the Wisconsin Glacier crept to within forty miles of the northern edge of today's Pine Barrens, forcing tundra species such as broom crowberry, *Corema conradii,* to move south. According to pollen layers from that time, both the pitch pine and other pines of the boreal forest such as jack pine, *Pinus banksiana,* and red pine, *P. resinosa,* were present in the barrens. After the glaciers receded and the climate warmed once again, the oaks and southern pines reinvaded. Positioned as it was on the leading edge of shifting climates, the Pine Barrens thus acquired its unique mixture of northern and southern plants.

Because vast quantities of the earth's surface water froze into the glaciers, much of the continental shelf was exposed during that time. Botanists theorize that plants may have migrated north across this land, their seeds carried by swirling sou'westers born of subtropical air masses. Many species that colonized the pines during this period apparently became isolated, for some species there today grow nowhere else on earth. Others are disjunct populations, meaning they are not found again for hundreds of miles.

The soil of the Pine Barrens is best described as dirty sand. It is a northern podzol soil, out of place in New Jersey, a soil so sandy and porous that it swallows water immediately, leaching out the nutrients and leaving the surface as dry as a dune. Everywhere, the woods floor is covered with a litter of undecomposed oak leaves, pine needles, and cones, because with a pH of 4.0, the soil is far too acid for earthworms or bacteria. Between fires, three to four inches of this litter may accumulate on the forest floor, while the laurel, blueberries, and other shrubs grow into a tangled, resinous, inflammable mass several feet high. In summer, droughts are common, and a spark landing on dry litter will cause the waxy leaves of laurel and heaths to flare like tinder, igniting the pines above them into crackling torches.

Thus the fire and the soil have selected the vegetation of the Pine Barrens, favoring plants that are both highly inflammable and yet able to recover quickly after being burned. The pitch and shortleaf pines have thick bark that protects them from extreme heat, and within a surprisingly short time, dormant buds will sprout from blackened trunks and limbs and new green shoots will rise from the roots. The three typical oaks of frequently burned areas, the post, the bear, and the blackjack, are all fire resistant to varying degrees, yet only the blackjack oak, *Q. marilandica*, seems to actually thrive on fire and will come back even after the oldest pines have been killed. Then for a time it becomes the predominant tree of the woods, while all around it, germinating pine seeds sprout from the charred earth.

Two races of pitch pine have evolved over the millennia. The trees of the first race produce cones that open when they are ripe in the normal manner of most conifers, but the cones produced by the second race remain closed, sometimes for years, until they are exposed to the searing heat of fire. Both kinds occur throughout the Pine Barrens, yet all the pines in the dwarf forest of the Plains belong to the second race.

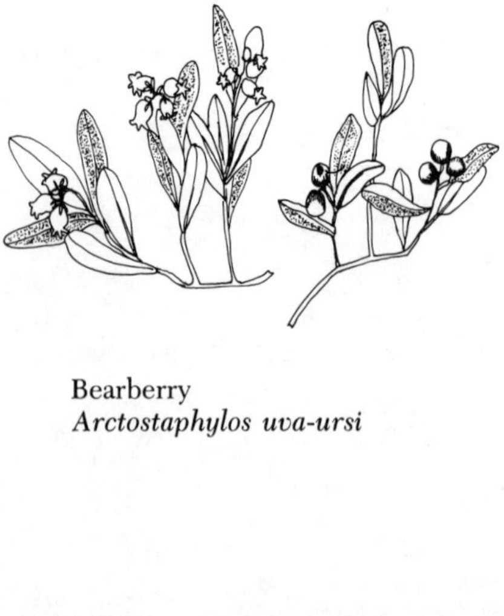

Bearberry
Arctostaphylos uva-ursi

Like the plants of the dunes, many plants that

grow on sandy, acid soil tend to form tufts or mats. Bearberry, *Arctostaphylos uva-ursi,* that plant of dune and tundra, also grows here. The diminutive pyxie moss, *Pyxidanthera barbulata,* is a tiny alpine plant from the Appalachians that forms tufted cushions on the sand, sending out radial branches that root in the undecomposed litter. In April, pyxie moss is covered with round pink or white flowers, each with a fused corolla that splits into five triangular petals. In the pines, it often grows with trailing arbutus, *Epigaea repens,* another creeping evergreen, and sweet fern, *Comptonia peregrina.* Sweet fern is not really a fern, but a shrub that forms dense thickets three to four feet high in burned-over areas. The Victorians grew it in their gardens for the pungent fragrance of its leaves. Bracken, *Pteridium aquilinum,* is a true fern of worldwide distribution with a root system so deep it is unaffected by fire. It too is common in the pines.

Trailing Arbutus
Epigaea repens

Sand myrtle, *Leiophyllum buxifolium,* is another mat-former that is found only in the Pine Barrens: a diminutive tuft of upright branches with minute leaves and pink or white flowers in loose terminal clusters. It often grows with *Hudsonia ericoides,* a close relation of the beach heather, *H. tomentosa.* This species is greener. It looks less like juniper and more like a club moss, except for its tiny yellow flowers.

The most widespread and conspicuous shrubs of the Pine Barrens, in both wet and dry habitats, are the heaths, family Ericaceae. In dry, sandy areas, black huckleberry, *Gaylussacia baccata,* and lowbush blueberry, *Vaccinium vacillans,* form an almost continuous knee-high blanket beneath the pines and the oaks. *Gaylussacia irondosa,* the dangleberry, is also common. To tell a blueberry from a huckleberry, look for small resin glands, like amber dots, on the underside of the leaves. If present, the plant is a huckleberry.

The laurels also grow everywhere, both mountain laurel, *Kalmia latifolia*, and the smaller, purple-pink sheep laurel, *K. angustifolia*, a plant that was once destroyed over much of its range because it is poisonous to cattle and sheep.

Turkeybeard, *Xerophyllum asphodeloides*, is a distinctive plant of the Pine Barrens and a member of the legendary bunchlily family. It blooms in June on soil that is dry on the surface but moist beneath, a condition not uncommon in the pines, where aquifers frequently approach the surface. Turkeybeard is a stately plant with spires of white flowers on sturdy three-foot stems rising from a hummock of grasslike leaves. It is a southern plant that originally migrated from the Appalachians onto the Coastal Plain and reaches its northernmost limit in the Pine Barrens.

Turkeybeard
Xerophyllum asphodeloides

Pink catchfly, *Silene caroliniana*, forms a tidy clump of flowers every spring. It is a member of the pink family, Caryophyllaceae, and pretty enough to grow in any garden. *Chrysopsis mariana*, the Maryland golden aster, is a handsome yellow composite with yellow rays and yellow disks that blooms on a stout, silky stem about two feet tall. It flowers in the Pine Barrens during August and September, and while it looks like a yellow aster, it is botanically closer to the goldenrods. The genus name translates as "golden face."

In the fall the silvery aster, *Aster concolor*, a species with conspicuous fuzz on its leaves and stem, blooms in the pines. It is listed as an endangered plant. More common but no less beautiful is *Aster spectabilis*, a plant that grows in the barrens with bearberry and sand myrtle. Last of all is the lovely Pine Barrens gentian, *Gentiana autumnalis*, also called one-flowered gentian. It is a slender plant, about a foot and a half tall with very narrow, delicate leaves. The color is usually the typical deep blue of the gentians, but sometimes an isolated plant may be white or purple.

Serpentine Barrens

Even before the land was settled there were clearings in the forest like miniature grassy prairies. Some of these ran in an irregular line from Philadelphia to the Susquehanna River that when viewed from above must have broken the sea of trees as a serpent's coils might smooth the surface of the water. These misplaced bits of prairie were called the serpentine barrens, named not for my metaphor, but for the peculiar green rock that underlays them. This same rock occurs in northern Italy, where it harbors a small snake of the same color.

Serpentine barrens occur in many parts of the world. Besides Italy, they are found in Cuba, Japan, and Norway. Some geologists think they may mark boundaries where tectonic plates ground together and forced molten serpentine through the surrounding bedrock.

Because the soil is thin over such rock outcroppings, it dries quickly, so dryness is one factor selecting plants for these sites. Yet it is not the only one, for the soil over serpentine lacks phosphorus, nitrogen, and calcium, all nutrients necessary for the health of most plants, and the rock also gives off a high concentration of magnesium ions as it decomposes. Magnesium is poisonous to some plants and apparently acts to dwarf others. Such a highly specialized habitat selects specialized plants, so it is not surprising that these barrens support some of our rarest American natives.

The most familiar rare plant of the barrens is moss phlox, *Phlox subulata*, which in early spring paints hundreds of eastern lawns and garden walls a vivid shade of hot pink. In the wild, the pink is far less strident, yet it is the reason that locally many serpentine outcrops are called "pink hill." The phlox often coats the steepest part, growing in a sort of green scree with another serpentine endemic, *Ce-*

Moss Phlox
Phlox subulata

rastium arvense var. *villosum*, a particularly handsome species of the lowly chickweed. The word *villosum* means "covered with soft hairs" and alludes to the fuzzy gray-green foliage. The white flowers are the size of a dime and have the notched petals typical of chickweeds.

The slender tufts of rock sandwort, *Arenaria stricta*, grow on these barrens, and a relative of portulaca called sunbright, or fame flower, *Talinum teretifolium*, a plant with wiry stems and pink flowers that last only one day. In summer the serpentine aster, *Aster depauperatus*, blooms, a bushy species that is covered with hundreds of tiny white flowers. Whorled milkweed, *Asclepias verticillata*, has long, slender leaves with the texture of rosemary. They are arranged in whorls around the stem, and in summer creamy white flowers bloom in the axils. Serpentine species of familiar genera often seem dwarfed, and this is true of the dry land snakeroot, *Eupatorium aromaticum*, a smaller, more delicate version of the familiar snakeroot of our autumn woods.

In recent years development has destroyed many serpentine barrens, and pollution and acid rain have altered the soil of others, allowing multiflora rose and black locust trees to crowd out the indigenous flora. Burning apparently controls this to some degree.

Hoary Puccoon
Lithospermum canescens

Limestone Barrens

Outcrops of limestone, a gray and white sedimentary rock composed mainly of calcite, produce soils of a neutral pH unusual in the eastern woods. While neutral soils are extremely fertile, on these rocks the layer is too thin to retain moisture and so must be considered as a xeric or dry environment. Hoary puccoon or Indian paint, *Lithospermum canescens*, is a plant of the midwestern prairie that sometimes occurs on limestone barrens in the East. It forms a showy clump of oblong leaves that are covered with hairy fuzz and support brilliant orange-yellow flowers. The word "puccoon" was given by the Indians to many plants that were used for dye. Cliff green, *Pachistima canbyii*, also occurs on limestone barrens. It is a low-growing evergreen shrub much valued by gardeners as a ground cover.

Shale Barrens

Some Appalachian ridges have slopes that are thickly strewn with slabs and flakes of gray-brown shale. Geologically these outcrops date to the middle Devonian period, the age of the fishes, and date from about three hundred and twenty million years ago. They occur most often in Virginia and West Virginia, but there are some examples in western Pennsylvania as well. Shale barrens support some of the rarest plants of the East, including mountain clover, *Trifolium virginicum*, and shale morning-glory, *Convolvulus purshianus*. *Aster oblongifolius* is a rather stiff, bushy plant from the Midwest that sometimes blooms on shale barrens. Its flowers are lavender and among the more handsome of the aster tribe.

III
THE NAMES

9
WHAT'S IN A NAME?

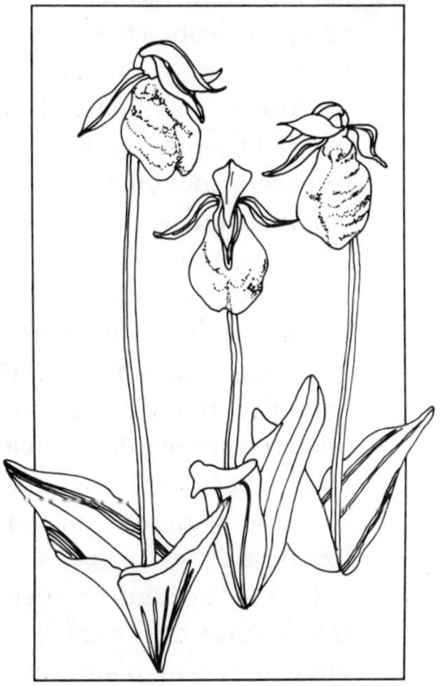

The common names of plants are filled with a droll sense of imagery. Unsung country poets of an earlier day named them after their animals, like goosefoot or pussytoes, after their clothes, like monkshood or moccasin flower, or simply for their dreams, like forget-me-not or live-forever. Names such as poor man's weather glass or beggar's ticks show a charming self-deprecating humor. Hercules'-club and Venus' looking glass show a feel for classical myth, while St. Andrew's cross or Solomon's seal indicate at least a nodding acquaintance with religious history.

Country people listened to plants, they smelled them, they tasted and touched them; and they named them rattlebox, bittercress, skunk cabbage or mouse-ear. They knew plants intimately, at all times of the day and at all seasons. For instance, St. Andrew's cross comes from the shape of the four-petaled flower, but it is the flat, polished seeds that form the "looking glass" of Venus, and the "seal" of Solomon only appears when the dead stem separates from the root.

These names are fun to know and use, but the truth is only a tiny fraction of the world's plants have common names at all, and most of those are only dry translations from the Latin. Greenish-flowered pyrola, or panicled aster, are the kinds of names bestowed by botanists. They simply can't compare with the imagery of prince's feather or lizard's tail, yet prince's feather or lizard's tail will not do for accurate research. Such names are usually restricted to a small area, or they may refer to quite different plants in different places. For instance, bush honeysuckle in the South means *Rhododendron*; in the North it is *Lonicera*. Fetterbush may mean either *Leucothoe* or *Pieris,* while bluebells can be *Mertensia*, *Polemonium*, or *Campanula*. Some common names can be frankly misleading: cypress spurge is no relation to cypress and red cedar is not a cedar but a juniper.

Latin is the language of scholarship, and plants have been named and described in Latin for centuries, often in beautifully illustrated volumes called herbals. Because plants were medicine, proper identification was vital; each had to be described accurately and in minute detail. A widely quoted example from one of these early herbals is a description of clove pink, called "gillyflower" in early English gardens. It reads: "Dianthus floribus solitaris, squamis calycinis subovatis brevissimus, corollis crenatis," and translates as "Dianthus with solitary flowers, with very short inverted egg-shaped scaled calyces, and crown shaped (crenelated) corollas." Detailed and accurate as this description is, it is undeniably awkward. Nor does it connect this dianthus to any other species or plant.

Botanical Nomenclature

Carl Linnaeus (1707–1778) not only loved plants, he was fascinated by their natural history and distribution. He wanted to give them names that would accurately describe a particular individual and at the same time connect it with other species in an orderly and expandable system. This in fact was his astonishing achievement. By the end of his life he had not only organized all the known plants, but the animals and minerals, too, in a way that made them accessible to scientists for the first time. Hundreds of thousands of new species have been discovered and named since his time, yet his system of nomenclature continues to work.

Linnaeus was not the first to try and simplify unwieldy descriptions. The French botanist Charles L'Ecluse (Latinized to Carolus Clusius) was giving plants two names as early as 1576, but there was no consistent system to his effort. Linnaeus was the first to see clearly that there were groups of plants that had certain characteristics in common. He called these groups *genera*. Each group contained many different kinds of plants that while resembling each other in certain basic ways, differed in other ways that were consistently reproduced. He called these individuals *species*. In his book *Species Plantarum* he assigned each plant a number and described it first in the traditional way in Latin. He then took the genus name and combined it with a short name that he called the *nomen triviale*, or trivial name. Technically it is now called the specific epithet, although it is usually referred to as simply the specific name. Ever since, these generic and specific names have always been used together and comprise what botanists call the binomial system of nomenclature.

The First International Botanical Congress met in Paris in 1867 in an effort to formalize Linnaeus's system so that it could be used by botanists worldwide. This meeting began an often stormy process that eventually led to the adoption of The International Code of Botanical Nomenclature, a code that has now standardized the naming of plants all over the world.

The most important parts of this code, from the point of view of the amateur, are as follows:

1. Valid publication of all botanical names began in May 1753 with the publication of *Species Plantarum*. This ruling formally acknowledged Linnaeus as the originator of the binomial system and validated all his original names.

2. Botanical names were decreed to be independent of zoological names. This rule gave scientists the freedom to use a particular generic name for animals as well as plants.

3. Each scientific name is to be followed by the name of the author. When correctly written, the name *Liriodendron tulipifera* should appear *Liriodendron tulipifera*, L., indicating that this plant was named by Linnaeus. If, as often happens, a

later botanist should reassign the plant to a different genus, the name of the original author is retained in parenthesis, as in *Ipomoea purpurea* (L.) Roth. Including the name of the author is a sort of botanical bookkeeping. It increases accuracy, but because it is clumsy for everyday use, it is usually omitted from less technical books.

4. A herbarium specimen of the original plant found by the author is always the standard against which all subsequent plants of the same species are compared. It is the nomenclature type for that species.

5. Publishing a new name requires certain formalities, but once published a name can only be replaced if a previously published name is discovered for that species. For instance, the name for pinxterbloom was recently changed from *Rhododendron nudiflorum* to *R. periclymenoides* when the latter was found to date from an earlier time.

6. Family names are formed by adding -aceae to the stem of a prominant genus within the family. There are eight exceptions to this: family names that were in general use before the adoption of the rules. However, each exception was given an alternative name that complies with the code. The eight, with their alternate names, are as follows:

Compositae – Asteraceae	Composite or aster family
Cruciferae – Brassicaceae	Mustard family
Gramineae – Poaceae	Grass family
Guttiferae – Clusiaceae	St. Johnswort family
Labiatae – Lamiaceae	Mint family
Leguminosae – Fabaceae	Bean family
Palmae – Arecaceae	Palm family
Umbelliferae – Apiaceae	Parsley or carrot family

Older, more traditional botanists tend to stay with the old names, while younger botanists have often been taught the new ones exclusively.

Latin names, like common ones, tell their own stories, for they are full of fascinating allusions for those equipped to read them. The early naturalists, including Linnaeus, were fond of naming genera after fellow botanists or other people they admired. For instance, *Kalmia* was named for Peter Kalm and *Franklinia* for Benjamin Franklin.

Often generic names were taken from ancient Greek names, such as *Agrostis* for grass and *Fagus* for beech. Others are simply descriptive: *Sanguinaria*, the generic name for bloodroot, comes from the latin *sanguis*, meaning blood; *Daucus* is the Latin word for carrot; and *Chrysanthemum* comes from two Greek words: *chrysos* meaning gold and *anthos* meaning flower. Some generic names allude to the medicinal use of the plant. Examples would be *Schrophularia* which was thought to cure scrofula, or *Salvia* which means simply "to heal."

Specific names are more apt to be genuine Latin as well as more descriptive of the particular plant. Many refer to a plant's distribution: the first North American plants to be named were usually called *americana*. In an effort to be more specific the early botanists then named everything north of New York *canadensis*, everything from the Middle Atlantic states *pensylvanica* or *virginica*, and plants from the South *caroliniana*. These names are repeated over and over. As time went on, they became more specific, naming *Rhododendron catawbiense* for the Catawba River in South Carolina and *Franklinia alatamaha* for the Alatamaha River in Georgia.

Specific names also provide clues to the plant's appearance. For instance, *latifolia* means wide leaf and *angustifolia* means narrow leaf. *Tomentosus* means hairy, *giganteus* large, and *cernuus* nodding. Names that refer to a plant's habit of growth include *scandens*, which means climbing, or *repens*. which means creeping. Still other names refer to a plant's habitat: *palustris* means marsh-loving, *pratense* means of the meadows, and *arenarius* means growing in sandy places.

The rules of Latin grammar are loosely followed; that is, generic names always have a masculine (-us), feminine (-a), or neuter (-um) ending, and the ending of the modifying specific name follows suit.

People are often uncertain as to how to pronounce Latin names. It is simpler than it seems, for most American botanists simply pronounce the words as if they were English. All syllables are pronounced and the accent is always on the next to last syllable or the one directly before it, that is, the penult or the antepenult. If the word has only two syllables, pronounce the first. The last syllable is never accented.

Classification

The first man to try and arrange plants into an orderly system was not Linnaeus, but Theophrastus (370–285 B.C.), a student of Aristotle. He grouped them by form, separating trees, shrubs, and herbs, and differentiating between annuals and perennials. He even described the various structures of flowers, noticing whether their petals were fused or separate and if their ovaries were inferior (below the point where the petals are attached) or superior (above the point of attachment).

After Theophrastus, no further effort was made at classification until a French botanist, Joseph Pitton Tournefort (1656–1708), published a book called *Rei Herbariae* in 1700. Tournefort was the first to describe plants from an objective interest as well as for their medicinal value. He described over ten thousand plants, many beautifully illustrated. Like Theophrastus, he grouped them by form, but he also recognized incipient genera such as oaks, peas, and willows, and he separated flowers that had petals from those that did not. Tournefort's system, although still rudimentary by our standards, was widely used in Europe for nearly a century.

An English botanist called John Ray (1628–1705) was a contemporary of Tournefort. He devised the first system to be based on the structure of plants, that is, the characteristics of their fruits, flowers, and leaves, and he was also the first to recognize the difference between monocots and dicots. Although botanically superior, his system was overshadowed by Tournefort's and went largely unnoticed until long after his death.

Linnaeus had a passion for order and arrangement. His botanical system was numerical, based on the sexual organs of the flower. That is, he grouped plants primarily by the number of stamens, or male organs, dividing them further according to the stamens' length and structure. He also put plants with imperfect flowers together (those with only male or only female parts) and those with infertile flowers, or no sexual organs.

The value of Linnaeus's system was in its simplicity. Anyone who could recognize and count the parts of a flower could immediately identify and classify an unknown plant. However, while it showed the natural relationships of many plants, it tended to separate some that obviously belonged together, so it is not surprising that it was soon challenged.

Two early challengers were Bernard de Jussieu (1699–1756) and his nephew Antoine Laurent de Jussieu (1748–1836), both members of a prominent French botanical family. Their system was also based on the botanical structure of plants, distinguished by the fact that it was the first to recognize thirteen classes of angiosperms, or plants with seeds contained in an ovary, categories that are still in use today.

All these early systems were based on the belief that a particular species never changed. At that time, all living things on earth were presumed to have been created in their present form in the space of a single moment far back in history. In the nineteenth century, Charles Darwin's theory of evolution as presented in *The Origin of Species* shattered this idea and in the process revolutionized all theories of classifying plants.

The basic tenet of Darwin's theory is that all species of plants or animals present on earth at this moment are descended from primitive ancestors. They attained their present form slowly, over thousands of generations, gradually becoming more and more adapted to a particular environment.

In plants, the tools of change and adaptation were hybridization and mutation. That is, today's plants are the result of hybridization between other, possibly quite different, species of long ago. Mutation, a term meaning a spontaneous change in the molecular structure of a gene that is inherited by the next generation, was probably responsible for abrupt changes. Of course, plants are still hybridizing and mutating today, but if the offspring so produced are not superior to their parents, they disappear, a fact which has undoubtedly been true from the beginning.

The theory of evolution spawned a whole new perspective on the classification of plants. So many new systems were developed incorporating the new theory that it is not surprising that no single one is used universally today. The differences between them usually center on what might have happened in the distant past that would indicate how large groups of plants are related. As a rule, botanists tend to be either "lumpers" or "splitters," disagreeing about what constitutes a separate species and what does not. This may be disappointing news to the layman looking for a definitive and universal authority, but it does tend to humanize the scientist who, however intelligent and highly trained, is still just an observer in a complicated world.

Lest the reader assume that there are no points of agreement, however, we should add that certain broad classifications are universally accepted, and that botanists only dispute about how certain plants should be allocated among them. The categories of classification proceed from the general to the specific, dividing like the forking branch of a tree. They may be briefly summarized as follows:

Kingdom Is the subject a plant or an animal? The answer may seem self-evident, yet it is not so obvious on the microscopic level, where some minute swimming creatures contain chlorophyll.

Division Is it a vascular or non-vascular plant? Vascular plants have stems, roots, and leaves and include most of the conspicuous plants of the earth. Algae, fungi, mosses, and liverworts are nonvascular plants. They have no stems, roots, or leaves.

Subdivision There are four subdivisions. Two divide nonvascular plants into the thallophytes, or algae and fungi, and the bryophytes, or the mosses and liverworts. Vascular plants are divided into the pteriodophytes, consisting of the ferns and their allies, and the spermatophytes, or the seed-bearing plants. We will continue our outline with the seed-bearing plants.

Class If it is a seed-bearing plant, is it a gymnosperm or an angiosperm? Gymnosperms include four to six orders (botanists differ on this point), of which the best known is the conifers, or cone-bearing evergreens. Conifers include pines, firs, cedars, cypresses, and araucarias. The name gymnosperm means "naked seed" and refers to the fact that the seeds do not develop within a fruiting structure such as an ovary. Gymnosperms do not have flowers.

Angiosperms have flowers, and their seeds develop within an ovary. They also differ in certain vegetative structures.

Subclass There are two subclasses of angiosperms: the monocotyledons and the dicotyledons.

Monocots usually have flower parts (sepals, petals, stamens and pistils) in multiples of threes, and their long, straplike leaves have parallel veins. Their roots are often bulbs, corms, or rhizomes. Familiar monocot families include the grasses, Gramineae, the lilies, Liliaceae, and the iris, Iridaceae. Plants such as orchids, Orchidaceae, sedges, Cyperaceae, arums, Araceae, and amaryllis, Amaryllidaceae, are also monocots.

Dicots have flower parts in multiples of four or five, their leaves are net-veined, and they usually have fibrous roots. Most flowering plant families are dicots.

Order Orders are groups of related families. That is, they have certain phyletically important characteristics in common.

Family Families are groups of related genera.

Genus Genera are groups of related species.

Species The species is the basic unit of classification, but oddly enough there is no universally accepted definition of what a species actually is! The best we can do is call it a kind of plant that differs in essential ways from other kinds of plants and can reproduce itself.

IV
PROJECTS

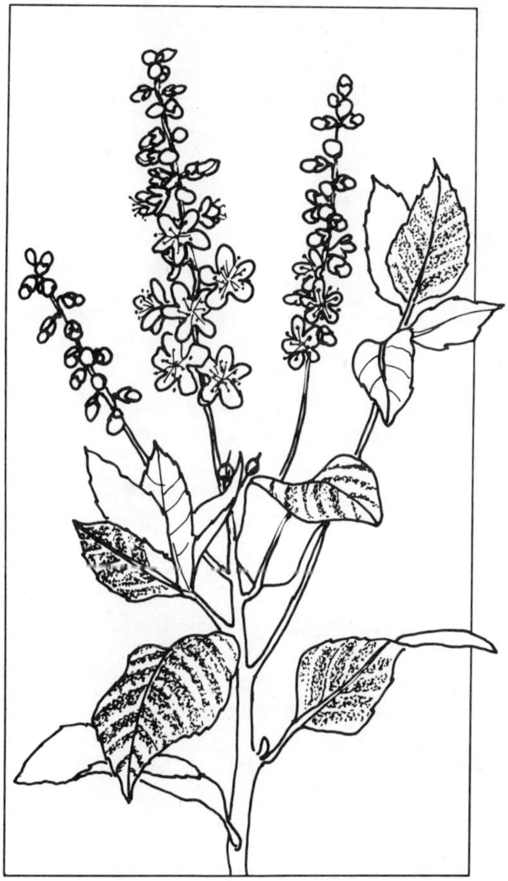

10
IDENTIFYING PLANTS

Sweet Pepperbush
Clethra alnifolia

Begin with a good field guide and a hand lens. The most popular lens is a small folding one that easily slips in your pocket or hangs on a cord around your neck. It magnifies ten times (10×), which is perfect for field work. A small sharp knife or razor is helpful, too, and a ruler, perferably one that has gradations in centimeters and millimeters as well as inches, because many botanical manuals use the metric system.

The Northeast

The best field guides for wildflowers of the northeastern states are *Newcomb's Wildflower Guide*, by Lawrence Newcomb, and Peterson and McKenny's *A Field Guide to Wildflowers*. Newcomb's identification system is preferred, because it teaches the beginner to observe the structure of the flower from the start. His guide also includes conspicuous flowering shrubs and vines.

In Peterson's guide, the flowers are grouped by color, a method that while easy and quick for the beginner, tends to encourage reliance on pictures to the detriment of botanical keys and descriptions. However, Peterson does have marginal symbols indicating the family of each species and a clear summary of family characteristics in the beginning of the book.

Definitely not a field guide, but nice to have for your own reference library, is The New York Botanical Garden's beautiful multivolume set called *Wildflowers of the United States* by Harold William Rickett. Two volumes are devoted to the wildflowers of the Northeast. They include clear descriptions of each species, but the photographs that illustrate them vary in quality. The work does not include grasses, ferns, trees, or shrubs.

For the rank beginner who wants to identify trees, a good place to start is with the Golden Guide *Trees of North America* by C. Frank Bockman. It is nontechnical and has attractive illustrations. More complete and more professional is Petrides's *A Field*

Guide to Trees and Shrubs. Thomas S. Elias's *The Complete Trees of North America* covers the entire country, has clear descriptions, and most important of all, range maps. It is an excellent book for the amateur but rather large to carry in the field. Sargent's *Manual of the Trees of North America,* vols. I and II, is thorough but extremely technical in its descriptions.

The best field guide for grasses and sedges is Lauren Brown's *Grasses. An Identification Guide.* The most complete work on grasses is A. S. Hitchcock's *Manual of Grasses of the United States.* For ferns, try *A Field Guide to the Ferns* by Boughton Cobb.

The definitive botanical manual for all plants found in the Northeast is H. Gleason's three-volume *The New Britton and Brown Illustrated Flora of Northeastern United States and Canada.* Dover Publications has reissued the second edition of this manual in three paperback volumes called *An Illustrated Flora of the Northern United States and Canada* by Nathaniel Lord Britton and Hon. Addison Brown, apparently because it covers a larger area than Gleason's revision. Yet, as this edition was originally published in 1913, it is too outdated to be of much use today.

M. L. Fernald's *Gray's Manual of Botany* is the other classic of northeastern flora. Unlike Britton and Brown, it is a single volume, but there are few illustrations.

The South

Two good field guides to southern wildflowers are *Wildflowers of the Southeastern United States* by Wilbur H. Duncan and Leonard E. Foote and *A Guide to Wildflowers of the Mid-South* by Arlo I. Smith. Unfortunately, they are both illustrated by photographs which vary in their clarity. Rickett's *Wildflowers of the United States* also has two volumes for the Southeast.

J. K. Small's *Flora of the Southeastern States* is the definitive manual for all species, herbaceous and woody, south of Virginia.

The Plains: West and Northwest

There are two good field guides for western wildflowers in the Peterson field guide series. They are: *A Field Guide to Rocky Mountain Wildflowers* by John J. Craighead, Frank C. Craighead, Jr., and Ray J. Davis, and *A Field Guide to Pacific State Wildflowers* by Theodore F. Niehaus.

The multivolume illustrated *Wildflowers of the United States* by Harold William Rickett mentioned previously also includes Texas (vol. 3), the Southwest (vol. 4), and the Northwest (vol. 5).

Among the more technical botanical manuals for the West are: *Wildflowers of the Pacific Northwest from Alaska to Northern California* by Lewis J. Clark, and two books by P. S. Rydberg: *Flora of the Rocky Mountains and Adjacent Plains*, 2nd edition, and *Flora of the Prairies and Plains of Central North America*. The latter was originally published in 1932 but has been reissued by Dover Publications.

Using a Botanical Key

There is no doubt that identifying an unknown plant can be daunting to the amateur. The plethora of obscure botanical terms causes many to give up before they have begun. For instance, there are over twenty terms for the shape of a leaf, nearly as many to describe its margins, and several more to describe its tip and its base. If a botanist wants to describe the hairy surface of a leaf, he has over sixty terms to choose from! No wonder so many give up in frustration.

Yet a botanical key is the most orderly of systems, proceeding one step at a time from the general to the specific characteristics of a plant. At each step you need choose only one of two possibilities. Your choice then leads to the next step, where you again choose between two possibilities, and so on until you have arrived at a precise identification. Botanical manuals usually have a key for plant families at the beginning of the book; then the section for each family has a key for the genera contained within it; and each genus a key for the species.

This is the theory. In practice, the choices are not so easy. If you should pick the wrong alternative, you will immediately veer off in the wrong direction, and the plant described at the end of your journey will not bear the slightest resemblance to the one you hold in your hand! Therefore, before you start, examine the plant carefully, so you will have a good idea of its characteristics. It is also important to have all the necessary pieces. The basal leaves of wildflowers often differ from those on the stem, and in the case of woody plants, the bud, twig, and leaf may be just as important as the flower and fruit.

It is impossible to know every botanical term by heart, but with practice you will soon acquire a good working knowledge of those most frequently used. An illustrated glossary is an enormous help: try James Payne Smith, Jr's. *Vascular Plant Families* or H. D. Harrington and L. W. Durrell's *How to Identify Plants.* For the serious student, the best book is George H. M. Lawrence's *An Introduction to Plant Taxonomy.*

As you proceed down the steps of the key, keep checking yourself. For instance, if your plant keys out to be a member of the mustard family, Cruciferae, read the description of this family to see if it fits. If you come to a place where both answers seem equally apt, mark it and proceed down each way in turn. When you come to a choice between two irrelevant descriptions, you will know you have taken the wrong road and can then go back and take the other branch. You may have to repeat this maneuver several times in a long key. Never try to go backwards; it simply won't work. Also, always check your final answer against the written description of the species.

If after all this the key still doesn't work, take heart: it may not be your fault. Keys vary in quality and usefulness. The author of the key might give ambiguous choices or fail to include all the alternatives, a fact particularly true of keys that try to cover large numbers of plants. They must concentrate on the average plant and cannot include every rare species or abnormal individual. Remember that large genera such as asters, *Aster*, and goldenrod, *Solidago*, have many species that hardly differ from one another. Others, such as hawthorn, *Cretaegus*, or shadbush, *Amelanchier*, tend to hybridize freely, often producing individuals that don't fit the textbook descriptions.

Amateurs tend to rely heavily on pictures, but they too can vary widely. Drawings are almost always better than photographs because the salient characteristics of the plant can be easily emphasized, but remember that "looking like the picture" is not enough for the professional. The species must conform to the written description before the identification can be considered accurate.

Parts of a Flower

Plants are sorted into families and genera according to the structure of their flowers, so a working knowledge of flower parts will make identification considerably easier.

If you look closely at a flower such as a single rose or apple blossom you will find that it has four rings of parts. The outer ring, called the *calyx*, is made up of the *sepals*. The next is the *corolla*, made up of the *petals*. Together, the calyx and the corolla form the *perianth*. If each petal and sepal is separate from the next, as in the rose, the flower is said to be *polypetalous* and *polysepalous*. If they are joined, as in a

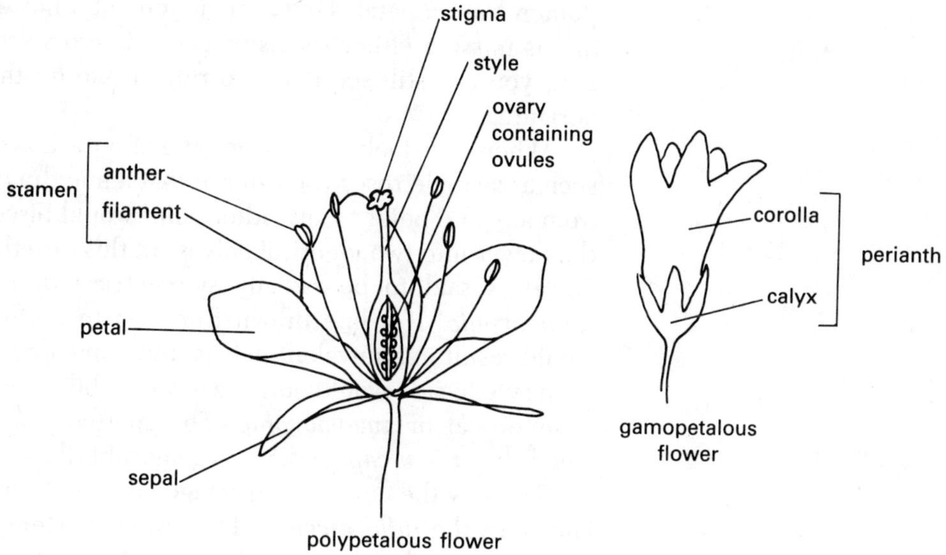

Parts of a Flower

morning glory, the flower is *gamopetalous* or *gamosepalous*.

The next ring within the perianth is the *androeceum*, literally the "male household." It consists of *stamens* (the male organs), each with a stalk called a *filament* and a pollen sac called an *anther*. The *gynoecium*, or "female household," is the fourth ring, found in the very center of the flower. The egg-containing organs, called *ovules*, are housed within an *ovary*. At the top of the ovary is a narrow neck called a *style* that is topped by the *stigma*. The flower becomes fertilized when pollen grains fall on the stigma, and the ovules then develop into seeds.

When all four of these rings are present, the flower is said to be "complete." It is "incomplete" if one or more rings are missing. A "perfect" flower has both male and female parts. An "imperfect" flower is all male or all female. If neither sex organ is present, the flower is sterile. All these parts come in a variety of shapes and sizes, sometimes imitating each other so closely that it is hard to tell a sepal or a stamen from a petal. However, if you cut a flower that is missing either sepals or petals in cross section, you can still see that two rings make up the perianth.

When you look down from above on a flower such as a single rose, you can see that a line drawn from any one point to any other point would bisect the flower into two identical halves. In this case the flower is said to be radially symmetrical or *actinomorphic*. If a line drawn from top to bottom would result in equivalent halves, but a line drawn from side to side would not, the flower is bilaterally symmetrical or *zygomorphic*. The spotted jewelweed, *Impatiens capensis*, is a zygomorphic flowers.

The way the flowers are arranged on the stem is known as the *inflorescence*. The flowering stem is called the *peduncle*. A peduncle with stemmed flowers arranged singly along it is called a *raceme*. A lobelia is a raceme. If the flowers arise directly from the peduncle without individual stems of their own, they are said to be *sessile* and the inflorescence is a *spike*. A *panicle* is a raceme with two or more flowers on each branch.

An *umbel* is a cluster of stalked flowers radiating from a single point. Queen Anne's lace, *Daucus carota*, is a compound umbel. A *head* is a dense cluster of tiny flowers called *florets*. In composites, there are two kinds of florets that make up the head: the disk florets in the center and the ray florets that surround it like petals.

Leaves

As mentioned in the beginning of this chapter, there are innumerable terms describing the shape of a leaf, as well as its margin, its tip, and its base. Still, it is helpful to know that some leaves are *simple*, like an elm, and some *compound*, like a hickory. The leaf of a hickory is *pinnately compound*, which means the individual leaflets are lined up along a central axis, as opposed to a buckeye leaf which is *palmately compound*: the leaflets arise from the base like the fingers of a hand. A leaf may also be *pinnately lobed*, like an oak, or *palmately lobed*, like a maple. Or it may be pinnately or palmately veined.

Woody plants are said to be *alternate* when the leaves and branches alternate along the stem or *opposite* when they are arranged in pairs. It is important to notice this arrangement when identifying trees and shrubs. Most woody plants are alternate. All those that are opposite are covered by the phrase MAD CAP, a handy assist for the memory that stands for Maple, Ash, Dogwood, and shrubs of the Caprifolaceae or honeysuckle family.

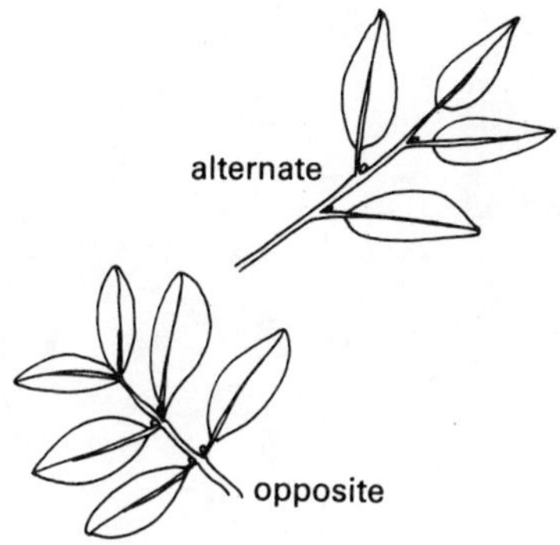

Arrangement of Leaves on a Stem

11
OTHER ACTIVITIES

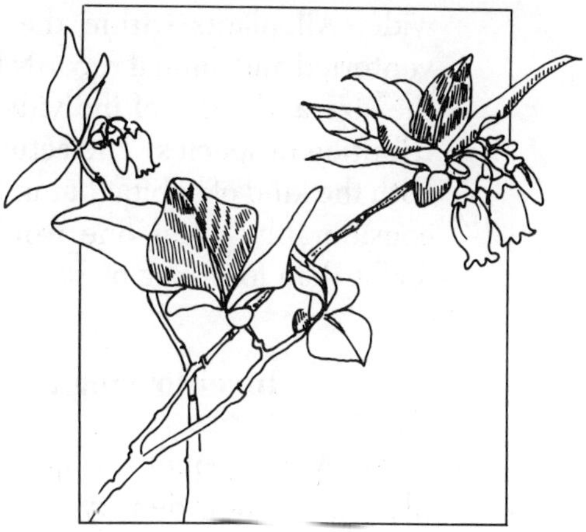

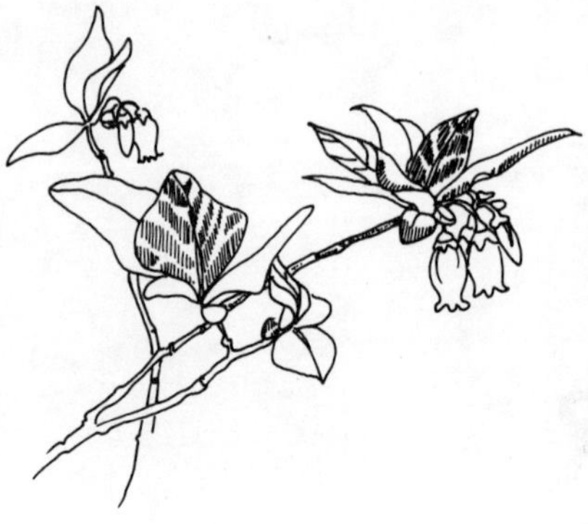

High Bush Blueberry
Vaccinium corymbosum

The enthusiastic amateur will want to keep records of observations in the field. If one is primarily interested in keeping a life list of the species one has found, a notation of date and location, recorded in the margins of a field guide, is usually sufficient. However, for readers interested in studying a particular plant community in detail, some discussion of methods may be useful.

A rough idea of a site's plant population may be had by inventorying points chosen at random or by walking a straight line from one end to the other and noting down every species as it is encountered. If the record is to span several years, however, it is a good idea to stake out a plot. The preferred shape is a long, thin rectangle as much as sixteen times long as wide. All plants within the plot can then be inventoried and annual records kept of any changes in the size and vigor of individuals as well as the distribution of species. The actual plot size would vary with the kind of habitat being studied; for instance, a considerably larger one would be necessary for a forest than for a bog or sand dune.

Inventorying a Sample Plot

1. As you count the species within your sample, notice their density as well as their variety. Does one species dominate all the others?

2. What is the relation between the individuals of one species: do they cluster together in a mass, or are they widespread?

3. How vigorous are the individuals? Do seedlings seem to die soon after germinating, or are they plentiful and widespread?

4. Does a particular life form, such as tap roots or succulents, predominant?

5. What are the systems of seed dispersal?

6. What is the rate of production of humus?

7. Notice the effects of light, heat, and moisture within the community. Notice, too, that the

majority of plants in a community often conform to a seasonal sequence; that is, most bloom either in spring or in summer.

Examples of Early Succession

1. Mark a sample plot in a field in the early stages of succession.
2. List all the species contained within it.
3. Note the numerical relationships between woody plants and herbaceous ones, as well as between grasses and forbs.
4. Note seedlings of new plants coming in. Are they the same or different from the dominants?
5. Repeat the experiment in other locations in the field. Are the results the same? If not, what do you think has caused the difference?
6. If left undisturbed, what do you think will happen in this field within the next five years? The next ten?

Examples of Late Succession

1. Mark another plot, this time within a young woods.
2. List species of canopy, subcanopy, shrub layer and forest floor, noting dominant species in each category.
3. Note saplings and tree seedlings. Are they the same species as the present dominants?
4. Note the presence or absence of woody vines and weak or dying trees. What species are they?
5. Estimate the approximate height and dbh (diameter breast high) of the largest trees.
6. What do you think this site looked like fifty years ago? If undisturbed, what will it look like fifty years from now?

7. Repeat the experiment at different sites of varying exposure and drainage. How do the results differ?

The early botanists grew trees, shrubs, and wildflowers from seeds and cuttings collected in the wild, and this is still the best way to collect wild plants without harming the native population. Growing from cuttings insures an exact replica of the parent plant. Growing from seed, while obviously much slower, is still the most economical way to get large quantities, and interesting variations often appear as well.

Propagating Wild Plants

Seeds

1. Seeds of trees and shrubs and wildflowers in the temperate zone must be stratified before they will germinate. This simply means they must have a period of cold to imitate the natural conditions of winter. An easy way to do this is to put the seeds in damp sphagnum moss and enclose them in a polyethylene bag. Label the bag carefully with the name of the species, the date, and location of the parent plant. Put it in the refrigerator for at least eight weeks or in the freezer for three. Some seeds, especially those with a hard coat, benefit from defrosting for twenty-four hours once a week and then refreezing, to further simulate natural conditions.

2. Seeds may be germinated in a commercial potting mix, fine-milled sphagnum, or even fine vermiculite, all of which are sterile and thus reduce the danger of fungus disease. If the plant's natural habitat is acid soil, milled sphagnum is best.

3. After the seeds have sprouted, feed them with a weak solution of water-soluble fertilizer. When two pairs of true leaves appear and they are

large enough to be handled, transplant them to large pots or flats. One part each of sharp sand, peat, and loam makes a good mix for many wild plants. If a more neutral or alkaline soil is needed, lime may be added.

4. Place the seedlings in the garden only after a period of gradual introduction to cooler temperatures. Meadow plants need sun. Woodland plants need good light, but bright sunlight will burn them. Keep the seedlings moist but not soggy. Protection with an overturned strawberry basket is helpful until they get acclimated.

5. Be patient. Biennials will not flower until the second year, and perennials will need two or three. Trees and shrubs are slow to start, but grow surprisingly fast after a year or two.

Cuttings

1. Timing is important when taking cuttings of woody trees or shrubs. Generally speaking, the best time is after the new growth has begun to harden off, usually about June or July, because then the plant has a relatively high stored food content.

2. Be sure your knife or pruning shears are clean and sharp to prevent injury and the transfer of disease. Make a slanted cut just below a node on the stem. Remove lower leaves, leaving about three or four leaves on the cutting. Particularly large leaves may be cut in half crosswise. It is important to keep the cutting turgid, so put it in a plastic bag if you are not ready to insert it in the medium.

3. A mixture of half perlite and half vermiculite makes a good rooting medium, although many people use sharp sand or a mixture of half peat and half sand. Fill a clean container, with a drainage hole, to within one-half inch of the top. Water well and drain so the medium is moist but not soggy.

4. Dip the cut surface only into a rooting hormone designated for use on woody plants. Do not immerse the entire stem. The hormone should only coat the slice.

5. Make a hole in the medium with a dibble or pencil and insert the cutting about one and a half inches deep. When doing several, keep them about three or four inches apart to maintain good air circulation. It is helpful to water the cuttings with a weak fungicidal solution.

6. Bury the pot to a half inch of its rim in the cold frame. If necessary, cover the frame to shade the cuttings, as sun will scorch the leaves and dry out the soil. Check them regularly to keep them evenly moist. As the cuttings will now stay in place for a full year, the frame should be insulated with matting over the winter. Deciduous cuttings drop their leaves and do not need light, but care must still be taken that they do not dry out.

7. Leave the rooted cuttings *in situ* through the next summer and plant them out in the fall.

Ways to Help Save Rare and Endangered Plants

1. Work to preserve habitats of rare plants by increasing public awareness.

2. Get involved politically. Even publicly owned land must have legislative restrictions on critical habitats in order to prevent public destruction. Because private land is particularly vulnerable to economic pressures, creative ways must be found to adequately compensate owners for saving land.

3. Help educate the public about the importance of plants to human welfare. Whenever a plant becomes needlessly extinct, we may lose valuable biological knowledge.

4. Support organizations that conduct field

surveys, germination trials, and life history studies. Get involved in these studies yourself.

5. Never dig plants from the wild. If you wish to grow native plants in your garden, propagate them from seeds and cuttings.

6. Never buy plants from a nursery that digs them from the wild. Patronize only those that propagate their own stock.

BIBLIOGRAPHY

Bailey, L. H. *How Plants Get Their Names.* New York: Dover Publications. 1963.

Billings, W. D. *Plants and the Ecosystem.* Belmont, Cal.: Wadsworth Publishing Co., 1965.

Braun, E. Lucy. *Deciduous Forest of Eastern North America,* New York: Hafner Press, 1950.

Brokaw, Howard, P., ed. *Wildlife and America,* Washington, D.C.: Council on Environmental Quality, 1978.

Brown, Lauren. *Grasses. An Identification Guide.* Boston: Houghton Mifflin, 1979.

Brown, Lauren: "Reediscovery." *Horticulture* 59, no. 2 (Feb. 1981) p. 32.

Bruce, Hal. *How To Grow Wildflowers, Wild Shrubs and Trees in Your Own Garden.* New York: Alfred A. Knopf, 1976.

Brush, Grace S. "An Environmental Analysis of Forest Patterns." *American Scientist,* 70, no. 1 (Jan.-Feb. 1981), p. 18.

Chase, Agnes, *First Book of Grasses,* Washington, D.C.: Smithsonian Institution, 1964.

Daiber, et al. *Atlas of Delaware's Wetland and Estuarine Resources.* Newark: Univ. of Delaware, 1976.

Daubenmire, Rexford. *Plant Communities.* New York: Harper & Row, 1968.

Duddington, C. L. *Evolution and Design and the Plant Kindgom.* New York: Thomas Y. Crowell, 1970.

Duncan, Wilbur H. and Foote, Leonard E. *Wildflowers of the Southeastern United States.* Athens, Ga.: Univ. of Georgia Press, 1975.

Elias, Thomas S. *The Complete Trees of North America.* New York: Van Nostrand Reinhold, 1980.

Farb, Peter. *The Face of North America.* New York: Harper & Row, 1963.

Favretti, Rudy J. *Early New England Gardens. 1620–1840.* Sturbridge, Mass.: Old Sturbridge Village Publication, 1962.

Fenneman, N. M. *The Physiography of Eastern United States.* New York: McGraw Hill, 1963.

Folkerts, George W. "The Gulf Coast Pitcher Plant Bogs." *American Scientist,* 70, no. 3 (May-June, 1982) p. 260.

Forman, Richard T. T., ed. *Pine Barrens: Ecosystems and Landscape.* New York: Academic Press, 1979.

Gleason, Henry A., and Cronquist, Arthur. *The Natural Geography of Plants.* New York: Columbia University Press, 1964.

Gleason, Henry A. *The New Britton and Brown Illustrated Flora of the Northeastern United States and Adjacent Canada.* 3 vols. New York Botanical Garden, New York: Hafner Publishing Co., 1952.

Godfrey, Michael A. *The Piedmont.* San Francisco: Sierra Club Books, 1981.

Greulach, Victor A. *Botany Made Simple.* New York: Doubleday, 1968.

Handlin, Oscar, ed. *This Was America.* Cambridge: Harvard University Press, 1969.

Harrington, H. D. and Durrell, L. W. *How To Identify Plants.* Chicago: The Swallow Press, 1957.

Harshberger, John W. *The Vegetation of the New Jersey Pine Barrens.* New York: Dover Publications, 1970.

Hitchcock, A. S. *A Manual of Grasses of the United States.* 2nd ed. rev. by Agnes Chase. Washington, D.C.: U.S. Government Printing Office, 1951.

Hodge, Walter Hendricks. "Where a Heavy Body is Likely to Sink." *Audubon Magazine,* 83, no. 5 (September, 1981) p. 98.

Howes, Paul Griswold. *Mini-Wood Community.* Hicksville, N.Y.: Sylvanus Books, 1975.

Jaynes, Richard A. "Chestnuts." *American Horticulturist,* 55, No. 6, (Winter 1976) p. 10.

Jorgenson, Neil. *Southern New England.* San Francisco: Sierra Club Books, 1978.

Kastner, Joseph. *A Species of Eternity.* New York: Alfred A. Knopf, 1977.

Kastner, Joseph. "Colonial Botanist, Self-Taught, Filled European Gardens," *Smithsonian Magazine,* 8, no. 7, (October, 1977) p. 122.

Kopper, Philip. *The Wild Edge*. New York: Penguin Books, 1979.

Lawrence, George H. M. *An Introduction to Plant Taxonomy*. New York: Macmillan, 1962.

Leet, L. Don, and Judson, Sheldon. *Physical Geology*. 3rd ed. Englewood Cliffs, N.J.: Prentice-Hall, 1965.

Leighton, Ann. *American Gardens in the Eighteenth Century*. Boston: Houghton Mifflin, 1976.

Lemon, James T. *The Best Poor Man's Country*. Baltimore: Johns Hopkins Press, 1972.

Leonard, Jonathan Norton. *Atlantic Beaches*. New York: Time-Life, 1972.

Madson, John. "Grandfather Country." *Audubon Magazine*, 84, no. 3, (May, 1982) p. 40.

Magee, Dennis W. *Freshwater Wetlands, A Guide to Common Indicator Plants of the Northeast*. Amherst: Univ. of Mass. Press, 1981.

McCormick, Jack. *The Life of the Forest*. New York: McGraw-Hill, 1966.

McPhee, John. *The Pine Barrens*. New York: Farrar Straus and Giroux, 1967.

Niering, William A. *The Life of the Marsh*. New York: McGraw-Hill, 1966.

Newcomb, Lawrence. *Newcomb's Wildflower Guide*. Boston: Little Brown, n.d.

Peterson, Roger Tory and McKenny, Margaret. *A Field Guide to Wildflowers*. Boston: Houghton Mifflin, 1968.

Petrides, Geo. A. *A Field Guide to Trees and Shrubs*. Boston: Houghton Mifflin, 1958.

Phillips, Claude E. *Wildflowers of Delaware and the Eastern Shore*. Hockessin: Delaware Nature Education Society, 1978.

Pyne, Stephen J. "Indian Fires," *Natural History*, 92, no. 2 (Feb., 1983) p. 6.

Rickett, Harold William. *Wildflowers of the United States. The Northeastern States*. 2 vols. New York: McGraw-Hill Book Co. 1966.

Sanderson, Ivan T. *The Continent We Live On*. New York: Random House, 1965.

Sargent, Charles Sprague. *Manual of the Trees of North America.* New York: Dover Publications, 1965.

Schaeffer, Elizabeth. "Researching a Pioneer Garden." *American Horticulturist* 59, no. 3 (June-July 1980) p. 24.

Schwartz, Susan. *Nature in the Northwest.* Englewood Cliffs, N.J.: Prentice-Hall, Inc., 1983.

Sculthorpe. *The Biology of Aquatic Vascular Plants.* London: Edward Arnold Publishers, 1967.

Sears, Paul B. *Lands Beyond the Forest.* Englewood Cliffs, N.J.: Prentice-Hall, 1969.

Silberhorn, Gene M. *Common Plant of the Mid-Atlantic Coast.* Baltimore: Johns Hopkins University Press, 1982.

Smith, Arlo. *A. Guide to Wildflowers of the Mid-South.* Memphis: Memphis State University Press, 1979.

Smith, A. W. *A Gardener's Book of Plant Names.* New York: Harper & Row, 1963.

Smith, James Payne, Jr. *Vascular Plant Families.* Eureka, Cal.: Mad River Press, 1977.

Snyder, Rachel. "Prairie Grasses for Gardeners." *American Horticulturist*, 57, no. 2 (April, 1978) p. 4.

Stevens, Larry. "King Kong Kudzu, Menace to the South." *Smithsonian Magazine*, 7, no. 9 (Dec. 1976) p. 93.

Stokes, Donald W. *A Natural History of Wild Shrubs and Vines.* New York: Harper & Row, 1981.

Stone, Hugh E. *A Flora of Chester County, Pennsylvania.* Philadelphia: Academy of Natural Sciences, 1945.

Teal, John and Mildred. *Life and Death of the Salt Marsh.* Boston: Atlantic, Little Brown, 1969.

Van Doren, Mark, ed. *Travels of William Bartram.* New York: Dover Publications, 1955.

Wheeler, W. A. *Forage and Pasture Crops.* New York: Van Nostrand, 1950.

Wilson, Charles Morrow. *Grass and People.* Gainesville: Univ. of Florida Press, 1961.

INDEX